Farming Systems
in the Nigerian Savanna

Westview Replica Editions

The concept of Westview Replica Editions is a response to the continuing crisis in academic and informational publishing. Library budgets for books have been severely curtailed. Ever larger portions of general library budgets are being diverted from the purchase of books and used for data banks, computers, micromedia, and other methods of information retrieval. Interlibrary loan structures further reduce the edition sizes required to satisfy the needs of the scholarly community. Economic pressures (particularly inflation and high interest rates) on the university presses and the few private scholarly publishing companies have severely limited the capacity of the industry to properly serve the academic and research communities. As a result, many manuscripts dealing with important subjects, often representing the highest level of scholarship, are no longer economically viable publishing projects--or, if accepted for publication, are typically subject to lead times ranging from one to three years.

Westview Replica Editions are our practical solution to the problem. We accept a manuscript in camera-ready form, typed according to our specifications, and move it immediately into the production process. As always, the selection criteria include the importance of the subject, the work's contribution to scholarship, and its insight, originality of thought, and excellence of exposition. The responsiblity for editing and proofreading lies with the author or sponsoring institution. We prepare chapter headings and display pages, file for copyright, and obtain Library of Congress Cataloging in Publication Data. A detailed manual contains simple instructions for preparing the final typescript, and our editorial staff is always available to answer questions.

The end result is a book printed on acid-free paper and bound in sturdy library-quality soft covers. We manufacture these books ourselves using equipment that does not require a lengthy make-ready process and that allows us to publish first editions of 300 to 600 copies and to reprint even smaller quantities as needed. Thus, we can produce Replica Editions quickly and can keep even very specialized books in print as long as there is a demand for them.

About the Book and Authors

Farming Systems in the Nigerian Savanna: Research and Strategies for Development
David W. Norman, Emmy B. Simmons, and Henry M. Hays

Presenting the case for a farming systems approach to research in developing countries, this book considers the role of new technology and appropriate development strategies in improving agricultural production and the welfare of farming families in the semi-arid tropical region of West Africa. The authors draw extensively on comprehensive studies they and their associates conducted over an eleven-year period in northern Nigeria. Their discussion of these studies, which focused on production, consumption, and marketing systems and included the testing of improved technology packages, is supplemented by results from research undertaken in other parts of semi-arid West Africa. Emphasizing the importance of a proper understanding of the technical and human environment in which farming families operate, they describe the essential characteristics of a farming systems approach and consider problems of methodology and implementation that must be solved if it is to become a widely accepted development strategy in the 1980s.

David W. Norman is professor of agricultural economics at Kansas State University. *Emmy B. Simmons* is an agricultural economist in the Program and Policy Coordination Bureau of the U.S. Agency for International Development in Washington, D.C. *Henry M. Hays* is associate professor and head of the Department of Economics and Business at Bethany College.

Farming Systems
in the Nigerian Savanna
Research and Strategies
for Development

David W. Norman,
Emmy B. Simmons, and
Henry M. Hays

Routledge
Taylor & Francis Group

LONDON AND NEW YORK

First published 1982 by Westview Press, Inc.

Published 2018 by Routledge
52 Vanderbilt Avenue, New York, NY 10017
2 Park Square, Milton Park, Abingdon, Oxon OX14 4RN

Routledge is an imprint of the Taylor & Francis Group, an informa business

Library of Congress Cataloging in Publication Data
Norman, D. W. (David W.)
    Farming systems in the Nigerian savanna.
    (A Westview replica edition)
    Bibliography: p.
    Includes indexes.
    1. Agricultural systems--Nigeria. 2. Agricultural research--Nigeria.
3. Agriculture and state--Nigeria. 4. Savannas--Nigeria. I. Simmons, Emmy B.
II. Hays, Henry Merlin. III. Title.
S473.N5N67 1982   338.1'09669   82-11077
ISBN 13: 978-0-367-02017-0 (hbk)
ISBN 13: 978-0-367-17004-2 (pbk)

To our families and
the farmers of the
Nigerian savanna

Contents

Tables

Figures

Maps

Acknowledgements

This book embodies work stretching over a fifteen year period. As a result inevitably we are greatly indebted to a number of institutions and a large number of individuals. Over the years considerable administrative and financial help was provided by the Institute for Agricultural Research, Ahmadu Bello University, the Ministries of Agriculture and Natural Resources in the northern states of Nigeria, the Ford Foundation, and more recently, Kansas State University.

With reference to individuals we would like to express particular appreciation for the encouragement and support, at different times, of Dr. I. S. Audu, former Vice Chancellor of Ahmadu Bello University, Drs. H. S. Darling and M. Dagg, former Directors of the Institute for Agricultural Research, Ahmadu Bello University, Drs. L. S. Hardin, W. Gamble and E. H. Gilbert, formerly of the Ford Foundation, Drs. J. B. Sjo and P. L. Kelley of Kansas State University and Dr. C. K. Eicher of Michigan State University. Special thanks are due to many former colleagues in the Institute for Agricultural Research and the Department of Agricultural Economics at Ahmadu Bello University. Senior staff colleagues who directly or indirectly contributed to the studies discussed in the book included: G. O. I. Abalu, C. Agbo, O. Aligbe, P. Beeden, L. Bungudu, B. J. Buntjer, B. D'Silva, N. O. O. Ejiga, R. J. Engelhard, E. Etuk, J. C. Fine, C. J. N. Gibbs, A. D. Goddard, H. R. Hallam, J. H. Hayward, B. Huizinga, A. H. Kassam, M. S. Krishnaswamy, W. J. Kroeker, N. B. Mijindadi, A. O. Ogungbile, J. O. Olukosi, S. I. Orewa, R. J. Palmer-Jones, D. H. Pryor, M. Tiffen and J. P. Voh. In addition many junior staff members within the Department of Agricultural Economics and Rural Sociology at Ahmadu Bello University helped in implementing the surveys, analyzing the data, and preparing the reports. Among the many who could be listed are the following: Sabo Abubakar, Mary Adebija, Julius Adeoye, Hamza Akwanga, Joseph Alabi, John Asaka, Danladi Jarma, Grace Michael, Rukayat Ojelade, Jeremiah Oji, John Oji, Muili Oladimeji, Stanley Onwuchekwe, Isa Sada, Adamu Umaru, and Adamu Yaro.

More recently valuable help has been provided at Kansas State University. Art Hobbs, John Sjo, and Carroll Hess made constructive comments on the manuscript. Lowell Brandner and Grace Muilenburg edited the final draft while Lois Langvardt, Janeé Roche, Susie Winkler, and Dawn Belknap typed many drafts of the manuscript.

xxii

Kelly Cunningham drew the figures while Susan Casement was responsible for the subject index. Finally Linda Norman spent many hours proofreading the final manuscript.

Perhaps the biggest debt of gratitude is owed to all the farming families who participated in the various studies and who accepted with equanimity our continual questioning. Over the years we learned a great deal from them and as a result our lives have been much enriched through having had the opportunity to spend so long a time in Nigeria.

D.W.N.
E.B.S.
H.M.H.

Acronyms

AAAS	American Association for the Advancement of Science
ABU	Ahmadu Bello University (Nigeria)
AID	Agency for International Development (USA)
AMIRA	Groupe de Recherche sur l'Amélioration des Methodes d'Investigation en Milieu Africain (France)
ASA	American Society of Agronomy (USA)
AVV	Autorité des Aménagements des Vallées des Volta (Upper Volta)
CATIE	Centro Agronomico Tropical de Investigacion y Ensenanza (Costa Rica)
CFDT	Compagnie Francaise pour le Développement des Fibres Textiles (France)
CIMMYT	Centro Internacional de Mejoramiento de Maiz y Trigo (Mexico)
CMDT	Compagnie Malienne de Développement des Textiles (Mali)
CNRA	Centre National de Recherches Agronomiques (Senegal)
CSNRD	Consortium for the Study of Nigerian Rural Development (USA)
DALST	Division for Agricultural and Livestock Services Training (Nigeria)
ECA	Economic Commission for Africa (Ethiopia)
EUCARPIA	Congress of the European Association for Research on Plant Breeding
FAO	Food and Agricultural Organization of the United Nations
FDA	Federal Department of Agriculture (Nigeria)
FMT	Farmer Managed Test
FSAR	Farming Systems Approach to Research
FSP	Farming Systems Perspective
FSR	Farming Systems Research
GCP	Guided Change Project (Nigeria)
HMSO	Her Majesty's Stationery Office (UK)
IAAE	International Association of Agricultural Economists
IADS	International Agricultural Development Service (USA)
IAR	Institute for Agricultural Research, Ahmadu Bello University (Nigeria)
IBRD	International Bank for Reconstruction and Development
ICRISAT	International Crops Research Institute for the Semi-Arid Tropics (India)
ICTA	Instituto de Cienca y Tecnologia Agricolas (Guatemala)

IDS	Institute of Development Studies (UK)
IEDES	Institut d'Etude Economique et Social (France)
IFAN	Institut Fondamental d'Afrique Noire (France)
IFPRI	International Food Policy Research Institute (USA)
IITA	International Institute of Tropical Agriculture (Nigeria)
INRA	Institut National de la Recherche Agronomique (France)
INSEE	Institut National de Statistiques et d'Etudes Economiques (France)
IRRI	International Rice Research Institute (Philippines)
ISRA	Institut Sénégalais de Recherches Agricoles (Senegal)
NAFPP	National Accelerated Food Production Program (Nigeria)
NISER	Nigerian Institute for Social and Economic Research (Nigeria)
OACV	L'Operation Arachide et Cultures Vivrières (Mali)
OLC	Overseas Liaison Committee, American Council of Education (USA)
ORD	Organisme Regional de Développement (Upper Volta)
ORSTOM	Office de la Recherche Scientifique et Technique Outre-Mer (France)
PANS	Pests, Articles, and News Summaries
RERU	Rural Economy Research Unit (Nigeria)
RMT	Researcher Managed Trial
SAED	Societe d'Aménagement et d'Equipement du Delta (Senegal)
USAID	United States Agency for International Development

1
Introduction: Farming Systems, Agricultural Research and Development Objectives

"The 'reason' why governments tend to introduce distortions that discriminate against agriculture is that internal policies generally favor the urban population at the expense of rural people in spite of the much greater size of the rural population," and because of "a shrinking from the complexity and difficulty of the task of developing agriculture ."
Schultz (1980) and Wilde (1967)

In the past two decades, Nigeria--with about eighty million people--has acquired the means to effect its transformation from a struggling Third World agricultural nation to an oil-rich exporting power searching for its place in history and in the ranks of more developed countries. It has had problems during the transition. Since gaining independence in 1960, the country has survived a civil war and moved from a loose federation of states to a federal entity of nineteen states (Map 1.1). It has also recently managed the transition from military to civilian rule.

Reliance on oil revenues with the government's directed emphasis on infrastructure, education, and industrialization has promoted significant growth in all sectors but agriculture. As shown in Table 1.1, Nigeria's Gross Domestic Product (GDP) has grown more than sixfold since independence--to a total value of ₦16,755 million in 1976.[1] The GDP growth rate improved slowly between 1960 and 1966. In 1960, the agriculture sector accounted for 64 percent of GDP and approximately 80 percent of the labor force employment. From 1966 through 1976, the period of rising oil exploration, GDP is estimated to have increased at a real annual rate of 8.5 percent, and GDP per capita at an annual rate of 6 percent. Per capita income rose to an estimated ₦252 in 1976 (Central Bank of Nigeria 1978). By 1974-75, agriculture accounted for only 21 percent of the GDP, a decline of 43 percentage points. The proportion of the labor force employed in agriculture had dropped to 64 percent (Federal Republic of Nigeria 1975). This dramatic decline in agriculture's share of GDP and the labor force stems in part from the increase in oil's importance and the labor force transformation. However, there is some evidence that total farm output has fallen in absolute terms.

MAP 1.1
Nigeria's Nineteen States

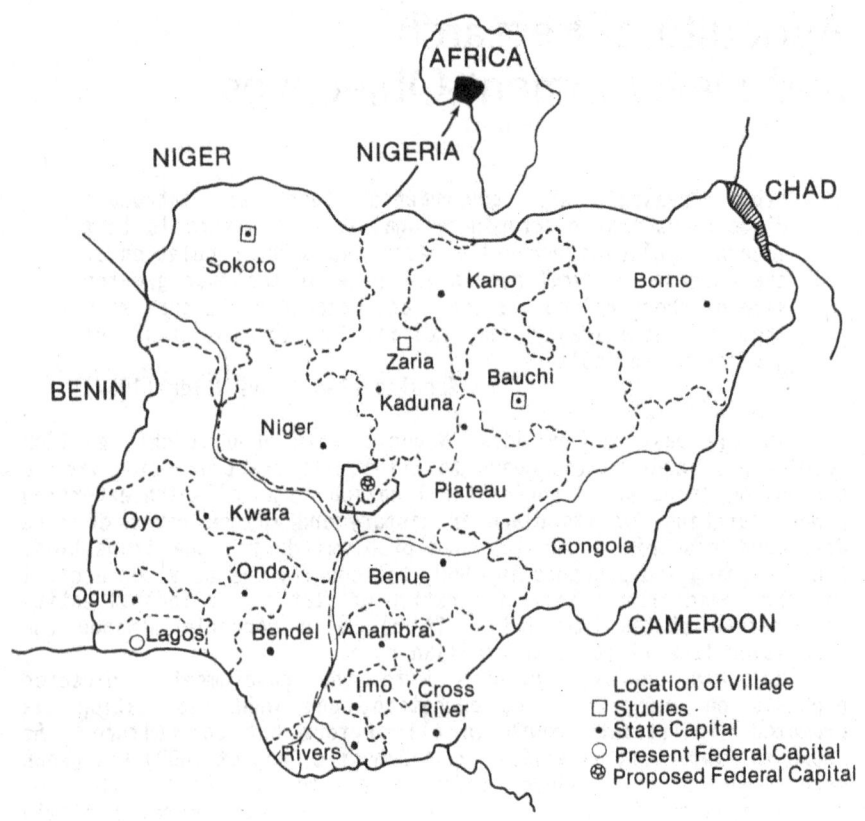

While these statistics probably reflect the adverse agricultural conditions of the early 1970's--the impact of the Sahelian drought on Nigerian agriculture--and somewhat overstate the decline, the impression of an agricultural sector lagging behind the rest of the economy is reinforced by both food import and agricultural export data for later years in the decade. Between 1973 and 1977, the food import bill rose sharply, from ₦126 million in 1973 to nearly ₦800 million just five years later. At the same time, agricultural exports fell to new lows. The value of the agricultural export index in 1960 was 100. Since 1970, the index has not exceeded 85 and in 1976, it plummeted to 68.

TABLE 1.1
Selected Performance Indicators of the Nigerian Economy

Item	1960	1966	1970	1976
The economy (million naira):[a]				
Gross domestic product	2493	3045	4178	16755
Agricultural output	1598	1582	1824	3491
Mining output	30	210	503	6886
Percent employed in the agriculture sector[b]	80	n.a.	n.a.	64
Indices:				
Production of major food crops	100	102	90	82
Consumer prices:				
All items	100	125	150	348
Food	100	133	164	465
Trade:				
Food imports	100	129	144	1102
Agricultural exports	100	115	101	68

Sources: Federal Office of Statistics (various issues); Federal
Republic of Nigeria (1975); Central Bank of
Nigeria (1977 and various other issues).

a. Figures for 1960, 1966, and 1970 were based on constant factor
cost for 1962-63 while 1976 was based on constant factor cost
for 1974-75.
b. n.a. means not available.

To talk of Nigeria's agricultural development thus involves
something of a misnomer. Production has declined, resulting in
greater disparities between rural and urban sectors and lack of
balanced development in the country. A more accurate description of
the past twenty years' experience might be agricultural
undevelopment.[2] But there is considerable concern about reversing
the trend (Essang 1978). Attention is being refocussed on the
agricultural sector and investments in various production activities
are beginning to support the rhetoric. Assuming that a realignment
of priorities for development will lead to further increases in
investment in Nigeria's agricultural sector, the question to be
answered is "what is the best way to increase productivity and
production with broad-based participation of all farmers and wide
impact in the rural sector?"
Nigeria is more fortunate than many developing countries in
having a substantial base of agricultural research infrastructure
and knowledge (Idachaba 1980) as well as financial and human
resources to use the knowledge. Still, Nigeria's leaders will
likely have to make some hard choices--which research gaps to fill,
which programs to support, which personnel to hire, which policies
to modify.
In this book, rather than offering definitive answers, we

suggest that starting with the farmers themselves is a useful way to begin. By adopting a farm-level- or micro-orientation, research problems relevant to changing the behavior of producers can be formulated and the research results, when achieved, can be more quickly fed back to stimulate production increases. By adopting a micro-orientation, extension programs can be adjusted to improve delivery of information and services relevant to client farmers. By adopting a micro-orientation, agricultural strategies and policies can be more closely geared to the incentive structures and resources of the producers themselves--with possible conflicts between societal goals and farmers' goals anticipated and ameliorated before bottlenecks become apparent and tensions arise.

More than eleven years of work at the Institute for Agricultural Research (IAR) in Zaria, Kaduna State, in the northern part of Nigeria, led us to this orientation. More recent work in other parts of Africa, Latin America, and Asia has persuaded us and others of the potential utility of such an approach in Nigeria and elsewhere. Assembling the factual base of empirical data needed to implement a micro-orientation is part of what already has come to be widely known as "farming systems research." Although a concise definition of what constitutes such research probably is not possible, the interdisciplinary approach and farmer involvement in research implied by the term are, we feel, critical to the development and application of a micro-orientation towards the problems of agricultural change.

Since our work in the Rural Economy Research Unit at the Institute for Agricultural Research helped support the emergence of farming systems research and our village-level research provides an early case study of its application, we present here both the theory and practice of farming systems research work. We attempt to place it in the context of agricultural research in general and agricultural development in the Nigerian savanna in particular.

In the remainder of this introductory chapter, we discuss in some depth the rationale for a micro-orientation to research and agricultural development activities and then briefly review the setting in Nigeria, where the micro-oriented research with which we were associated, evolved.

RATIONALE FOR A MICRO-ORIENTATION

In Nigeria, as elsewhere in the developing world, there has been an evolution in thinking about the problems of agricultural development. There has also been an evolution in thinking about how agricultural research might best be carried out to address development problems and goals. As would be expected, there are parallels between broad definitions of agricultural development approaches and delineation of agricultural research priorities and policies. Economic crises are increasing the pressures on developing countries to take a hard look at the dissemination of and return to government investments. Funds for agricultural research are not immune to such pressures. Where the returns from research do not seem commensurate with anticipated development impacts, governments often take steps to change the orientation of research to effect an improvement in the situation.

In the first section here, we trace the path which has led to the current concern with increasing the productivity of small farmers. In the second section, we discuss what this concern means in agricultural research terms: going back to basics and understanding the farmers.

Evolution of Agricultural Research Priorities

We believe that three or four decades ago, a dominant feature of agricultural research in developing countries involved satisfying the needs of the organization providing the research resources. These needs were not necessarily synonomous with the interests of farmers responsible for applying the technology.[3] In more recent periods, the thinking has shifted gradually to the view that the success of agricultural research must be measured in terms of its contribution to the welfare of the farmers themselves. The task of the agricultural research institution has thus become more complex. Not only is the research establishment responsible for executing a program consistent with national goals and scientific principles; it is also responsible for visibly improving the lives (and incomes) of farmers. The evolution in thinking can be broken into four stages.

In the first stage, the extractive philosophy of colonial times led to an agricultural development pattern concerned only with increased production of marketable surpluses for export (Lele 1975). The agricultural research emphasis was narrowly restricted to boosting the output of the export cash crops--in northern Nigeria's case, groundnuts and cotton. The colonial government ensured that research contributions were used by producers, but although some producers profited, benefits to producers were not a central concern.

In the second stage, the idea of selectively transferring technology to developing countries from developed countries supplanted the extractive approach. But the new approach was predicated on the notion that someone knew what was best for agriculture in a developing country. That resulted in attempts to import technology wholesale--sometimes with success but often with disastrous results. Heavy tractors became mired in mud, factories were installed to process ten times the volume of commodities available, dairy cows died of trypanosomiasis and other diseases. Where the wholesale transfer worked, dual agricultural economies often evolved, as, for example, in the case of Zambia. One, frequently nurtured and protected, became the modern sector of agricultural production; the other remained primitive and traditional (Norman 1981).

Then a third concept of developing agricultural technology within the low-income countries evolved. The unsuitability of directly-transferred technology contributed to this shift. The idea was that, by using as building blocks the elements that made technological change successful in high-income countries, researchers could develop unique and locally relevant technologies with a high degree of potential success.

In the fourth stage, those three essentially "top-down" approaches have been supplemented, but not entirely replaced, by a "grass-roots" or "bottom-up" strategy.[4] It is this latest stage of evolution that provides the foundation for the farming systems approach to research addressed here.

Among reasons for the changed thinking, perhaps the most fundamental was the repeated failure of other approaches to improve the lives and livelihoods of rural populations and in addition to meet the needs of the urban sector. Policies and technologies incompatible with the agroecological and the social, political, and economic environments were advocated (Hardin 1977). As a result, adoption rates were low except where compulsory measures were taken or where extraordinary input subsidies were extended, and the results expected did not materialize. A second reason is that where the well-being of rural populations was improved, neither the size nor distribution of benefits matched expectations.[5]

Although the production success of the Green Revolution should not be ignored, distributional problems engendered by such technology in South Asia, for example, have been widely portrayed as having led to worsening many farmers' positions vis-à-vis other farmers' achievements (Saint and Coward 1977). Despite claims that the seed-water-fertilizer technologies of the Green Revolution were intrinsically neutral to scale, the quality of resources required, together with differential access to the requisite infrastructural support systems, resulted in unequal benefits to farmers (Khan 1978; Poleman and Freebairn 1973; Valdes, Scobie, and Dillon 1979). Avoiding such inequities wherever possible is part of the renewed interest in the dynamics of rural development, so a prerequisite now is an income-generating force for the majority of farmers (Holdcroft 1978).[6] Unless agriculture is highly productive with a degree of market orientation, it will not generate the employment so significant in the new economics of growth propounded by Mellor (1976).

We contend that the checkered pattern of success will not be altered until the linkages among the three participants in the research process (sponsoring government or agency, research institution, and farmers) are strengthened and mutual accountability is increased. The top-down approaches, characterized by relatively tenuous linkages among participants, often have functioned poorly. Research institutes often have a difficult time communicating to the sponsoring agency what they have learned about the technologies, the farmers, and the research that should be done. Sponsoring agencies (normally governments, but donors supporting research with assistance funds also fall in this category) are relatively cut off from both research users (that is, the farmers) and the research institutes. Sponsors thus have problems translating their goals into action programs that are based upon realistic assumptions about the technology available, how it works, and how its adoption can be fostered. Unfortunately, sponsoring governments and research institutes are courted by different constituencies, not the farmers, so all research system participants are subject to conflicting demands for resources (Longhurst, Palmer-Jones, and Norman 1976).

Going Back to Basics: Understanding the Farmers

The quest for more effective ways of developing relevant improved agricultural technology and developmental strategies must,

therefore, involve the fourth, bottom-up, approach. It is analogous
to techniques used by commercial firms who measure their success in
sales; they first try to determine what their customers want and
then formulate a product to fulfill the want (demand). While
tailoring improved technologies to potential farmer-customers and
ultimately greater production are clearly the primary objectives of
bottom-up research strategies, there has been increasing recognition
that the farmers have something of value to contribute to the
development of technologies as well. Many practices currently used
by farmers, for example, are more fertility-conserving[7] and use
production factors more efficiently than some of the improved
practices. The realization that listening to farmers and observing
what they do can help to improve the potential for increased
efficiency in the allocation of research resources, emphasizes the
need for two-way communication between researchers and farmers.
Understanding farmers' methods of increasing soil fertility through
biological means, could, for example, be especially important as
researchers deal with the rising costs of fossil energy. While the
use of fuels and fertilizers is firmly embedded in much of the
improved agricultural technology developed to date, rising costs and
reduced availabilities are forcing researchers to re-examine
research priorities. Going back to basics--talking with farmers who
are effectively tapping other sources of fertility--is an
appropriate place to start.

INSTITUTIONAL HISTORY AND SETTING FOR MICRO-ORIENTED RESEARCH IN
NORTHERN NIGERIA

Background

     The Department of Agriculture in the former Northern Region of
Nigeria initiated technical research on agriculture problems in 1924
(Idachaba 1980). In 1957, research responsibility was transferred
to the Research and Specialist Division of the Ministry of
Agriculture of the Northern Region of Nigeria. When that Division
was transferred to the Ahmadu Bello University (ABU) in October
1962, the Institute for Agriculture and Special Services (IAR) was
established. By 1974, IAR had a senior staff establishment of 220
and an annual budget exceeding ₦3 million. It is now one of
eighteen agricultural institutes in the country covering crops,
livestock, fisheries, and forestry.
     The research mandate of IAR primarily covers the ten northern
states of Nigeria (Map 1.1), which include nearly 70 percent of the
total area of the nation and support approximately half of Nigeria's
total population. The academic mandate of IAR is to supply
researcher staff time to the Faculty of Agriculture at ABU. The
research arm of IAR includes six departments,[8] each subdivided into
two or more sections. To support the departments' academic
responsibilities, many department members have split appointments of
teaching and research. Such an arrangement, rather unique in Africa
(Bunting 1979), although common in United States land grant
institutions, permits the complementarity of teaching and research
to be exploited. The institutional structure is well suited to
permitting the use of the interdisciplinary farming systems approach

to research we advocate. At the same time it ensures that research funds available to academics are used to carry out research relevant to the region.[9]

The Institute for Agricultural Research also has a separate and distinct extension arm, the Division of Agricultural and Livestock Services Training (DALST).[10] It is a link between the research/academic staff and the extension workers employed by the Ministry of Natural Resources in each of the ten northern states.

Determination of Research Priorities

Several mechanisms have been developed to strengthen the link between government and research work at IAR that focusses on issues of immediate development concern. An annual meeting--known as the Cropping Scheme--is attended by most IAR researchers and representatives of all state Ministries of Agriculture and Natural Resources. It is the culmination of a process of review and discussion which involves a structured relationship between government and university representatives.

The IAR Professional and Academic Board places annual research proposals and reports prepared by IAR subcommittees before the Board of Governors composed primarily of senior agriculturists representing northern state governments and the Federal Government. The Professional and Academic Board, chaired by the Director of IAR, includes the Deputy Directors of IAR, the Department and Section Heads of IAR, the Provost of Agriculture, and elected staff representatives. The subcommittees of the Professional and Academic Board are formed along commodity lines and include staff members involved in research related to the commodities. Representatives of the Department of Agricultural Economics and Rural Sociology and DALST are generally represented on all subcommittees. The subcommittees initiate research plans and assess the suitability of proposed recommendations before the extension service disseminates them. The subcommittees' diverse memberships encourage an interdisciplinary approach to research problems.

Financial estimates of budgets needed to carry out the research programs agreed upon by the subcommittees are drawn up by the various departments of the IAR. Both research programs and financial requests are approved by the Professional and Academic Board before being transmitted to the Board of Governors for final approval.

The Socio-Economic Input to Agricultural Research

Socio-economic research in IAR is fairly recent. The first social scientists were appointed in 1965, forty-one years after technical agricultural research began in northern Nigeria,[11] with the creation of a new organizational unit. Initial financial support for this Rural Economy Research Unit (RERU) came from a Ford Foundation grant. By the early 1970s, however, most of the financial support came from Nigerian sources, and most of the research initiated by RERU was continued by the Department of Agricultural Economics and Rural Sociology.[12] By 1974, 10.5 percent of the research senior staff positions in IAR were in the social sciences, and social science research accounted for 8.3 percent of IAR's total research budget. Two factors appear to have influenced IAR to appoint its first social scientists. First, experimental yields of many crops were much higher than those obtained under normal farming conditions in the northern states and it was readily apparent that Nigerian farmers had adopted few improved technologies over the years, especially in food crops. Why? It was thought that a socio-economic approach could provide an explanation. Secondly, rural development programs in the northern states of Nigeria have, in general, emphasized voluntary participation rather than compulsion, so the idea of working with the farmers, usually small-scale ones, in traditional settings, rather than moving farmers to irrigation schemes, settlement schemes, etc. seemed sensible.[13] In such circumstances, emphasizing improvement rather than transformation-type strategies was felt likely to have greater payoffs. The addition of the socio-economic component, it was thought, could help provide information on types of improved technology acceptable to farmers.

It is apparent, however, that the implications and ramifications of the social science appointments did not immediately become apparent. Nor were social and technical scientists immediately integrated into a common effort. Although RERU was institutionally linked with IAR from the beginning, one of the first steps was to establish the credibility of that linkage--and of the social scientists involved--with other staff in IAR. A four phase research program was envisioned (Norman 1973a); the first two phases were embarked upon immediately by RERU but the third and fourth phases were undertaken only after elementary credibility and acceptance had been built between the technical scientists at IAR and the RERU staff.

The four phases in the research program were planned to provide a background against which relevant, improved technology could be designed and tested. The positive phase involved finding out what farmers were doing. The hypothesis-testing phase focussed on why farmers did things the way they did. Those two phases were expected to lead into the normative phase; that is, determining what ought to be done. The fourth phase of research activity involved policy and program analysis--determining what measures were required to accomplish the normative tasks, what ought to be done.[14] Examining incentives for farmers was an important part of phase four, as achieving development goals predicated upon farmers' participation is likely to depend on policies that provide appropriate incentives, when farmers are free to accept or reject agricultural innovations.

Much of the earlier RERU research work concentrated on the positive and hypothesis-testing phases. The so-called "basic studies" emphasis of those phases gradually shifted towards the "change studies" which have been associated with the normative and policy phases, although the linearity that implies is not a necessary element of the four-phase conceptualization.

The basic studies sought to describe, explain, and understand the rural/agricultural environment. Effort was on carrying out detailed village studies in five agroecologically distinct areas of the northern states. Interdisciplinary research work was accomplished among social science disciplines--geography, rural sociology, and agricultural economics--at certain phases of activity and not at others. Initial demographic and land utilization analysis for the villages usually was done cooperatively. While efforts were made to ensure that research by different disciplines fitted into the outline of the overall research program, researchers could pursue inquiries along disciplinary lines.

The change studies sought to assess the potential value of the agricultural technology being produced by technical research workers at IAR. Work constituting the change studies fell into three broad groups: first, assessing technical recommendations put out-- or scheduled to be released--by IAR; second, assessing governmental programs of agricultural change; and third, assessing different ways of introducing agricultural change.

Technical recommendations were evaluated for technical feasibility, economic profitability, and social acceptability. Emphasis was laid on investigations at the farmers' level rather than on experiment station results, usually dealing with one crop at a time. Investigations on cotton, maize, sorghum, cowpeas, and groundnut recommendations involved substantial interdisciplinary work between social and technical scientists.

Three government supported programs designed to bring about agricultural change were assessed: the Farm Training Institutes (Olukosi 1976), the Kadawa Irrigation Scheme, and the tomato-growing campaigns associated with establishing a tomato paste factory in Zaria (Agbonifo 1974; Orewa 1978). Results of the assessments are reported elsewhere so they are noted here only when they highlight particular points. Wilde (1967) has noted the tendency to repeat mistakes; such repetition is unavoidable unless past experiences are analyzed and recorded and the records are easily accessible.[15]

One major study, the guided change project, attempted to determine the best operational way to increase incomes from rainfed agriculture when faced with the administrative, financial, and manpower constraints normally experienced by state governments. The project, from knowledge accumulation through implementation, was made possible only by substantial cooperation between the North Central (later called Kaduna) State Government and IAR.

Much has been learned as a result of the various studies undertaken first by the Rural Economy Research Unit and later by the Department of Agricultural Economics and Rural Sociology. Much of the empirical work on which this book is based results from the basic studies and technology assessments. But it would not be true to say that the entire social science research program has been an unqualified success. Both methodological and administrative mistakes have been made, although even in retrospect, it is difficult to see that many could have been avoided. In several cases, avoiding mistakes would have involved knowing what the solution or response was likely to be before asking the question, clearly a difficult state of affairs to bring about!

Nevertheless, establishing a viable socio-economic research program in an agricultural research institution has taught us several useful lessons:

1. Placing considerable emphasis on a micro-oriented

approach--getting to know the farmers--is justified. Voluntary participation of farmers in programs of agricultural change and the predominance of farming systems that incorporated few, if any, modern practices dictate that researchers understand the problems and constraints farmers face before evolving types of improved technology and strategies that are designed to solve those problems and constraints.

2. Such farm-level understanding can be valuable in helping determine the research priorities of technical scientists and the design and implementation of macro-level agricultural development strategies that will be relevant to farmers who are expected to use them.

3. Development is a complex process so a multidisciplinary team working together in an interdisciplinary framework is more likely than a single-discipline approach to come up with research results that contribute to success. But it is true that interdisciplinary effort is difficult to achieve.

These three lessons will be expanded upon and illustrated as we describe the research efforts in greater detail. For, as the next chapter describes, the development and application of a farming systems approach to research forces researchers (both technical and social science) to examine and re-examine their work--not only in terms of publishable results acceptable to their disciplinary peers, but also in terms of contributions that can improve farmers' operations.

PLAN OF THE BOOK

A detailed description of the farming systems approach to research is provided in Chapter 2. A conceptual model delineates both the controllable and uncontrollable variables that impinge on the farmers' decision-making process. Using a farming system orientation we describe a method for designing relevant improved technology and developmental strategies. The conceptual framework provides the foundation for the empirically based discussion in the remaining chapters of the book which primarily concentrate on rainfed agriculture undertaken by settled farmers. Thus irrigation and nomadic livestock herding are not considered in detail although they of course contribute significantly to the economy of northern Nigeria and the savanna region of West Africa.

In Chapter 3 we describe the semi-arid climate and soils of the savanna region, which have major influences on the crop and animal ecology of the area. A brief review of agroecological characteristics provides a base for discussing the crops and livestock in the savanna region as well as the technical problems involved in raising them. At the end of the chapter we compare the potential yields of some of the major crops, as obtained under experimental conditions, with yields obtained under indigenous farming conditions.

In Chapter 4 the socio-economic organization of farming communities in the savanna region is described; particular attention is directed to interactions among the people within villages, compounds, and households. Some of the norms and beliefs, which help determine the transition of a society, are briefly enumerated. Linkages important to the attempts of governmental agricultural institutions promoting change and agricultural development are examined with a view to delineating the resulting influences and impact.

Chapter 5 is devoted to an empirical analysis of farming system determinants and draws upon research involving repeated interviewing and observation in three villages in the Zaria area of northern Nigeria during different periods between 1965-73. The data set, covering 124 farming households, resulted from major field surveys focusing on farm production, consumption, expenditure and marketing, credit, and storage. Various facets of income generation and employment are analyzed with a view to highlighting goals, constraints, and achievements of farm households. We conclude the chapter by assessing factors that determine the characteristics of farming systems and by giving special consideration to the implications of the assessment for improving household productivity and welfare.

Drawing upon empirical evidence obtained from village studies in the Sokoto and Bauchi areas of the Nigerian savanna, we analyze the factors underlying the diversity of farming systems in Chapter 6. The analysis includes references to studies in other parts of the West African savanna outside Nigeria, which permits a more complete consideration of the trends occurring over time in a fragile ecological zone together with a broad-ranging assessment of their implications for farmers in the savanna region. The considerations are important in highlighting critical issues in agricultural development, in helping formulate strategies for producing relevant technology, and in designing appropriate developmental policies.

In Chapter 7 we illustrate the testing phase of the farming systems approach to research, demonstrating the necessity of fully understanding the total farming system before developing suitable improved technologies applicable to small farmers. The analysis of both empirical results and the interpersonal experiences gained in the process of collaborative work involved in testing improved sorghum, maize, cowpea, and cotton packages are presented. The results of another effort to test appropriate support systems to facilitate farmers' adoption of improved technology packages are also summarized. The chapter concludes with a more general discussion of issues involved in incorporating on-farm testing into the farming systems approach to research.

In the concluding chapter, we discuss the roles which the farming systems approach to research can play in informing and supporting the design and implementation of strategies for promoting agricultural development. Agricultural strategies designed to address farm-level constraints directly and cost-effectively--the targeted approach of many development projects--can usefully adopt a micro orientation. Farming systems research projects can play an active role in the development and delivery of improved agricultural technologies which fit current farming systems. Organization, location, and methodologies of farming system research projects are important, however, and a few of the issues which need to be resolved at the various phases of the program are discussed. Where development strategies seek to promote agricultural growth by more indirect means--roads, prices, disease eradication, etc.--the farming systems perspective can help to assess the likely efficacy of such strategies in improving rural welfare. Finally, some implications for Nigerian policy makers are briefly discussed.

NOTES

1. One naira (₦1) equals approximately $1.60. One hundred kobo constitute ₦1.
2. Both Olayide et al. (1972) and a study produced by the International Food Policy Research Institute (1977) have predicted considerable increases in food deficits if production is not accelerated. Abalu (1978) has found the same with respect to sorghum and millet which are the key food crops in the savanna part of the country. Abalu and D'Silva (1980) have also addressed this problem.
3. We do not wish to suggest that the end result of such research always benefited the funding agency at the expense of farmers' welfare, and that this was an implicit objective of the former. However we would assert that in essence this sometimes did occur in practice.
4. "Bottom up" refers to the strategy of starting the research process at the farmers' level by first ascertaining their needs, and then using these needs to determine research priorities. This contrasts with earlier "top-down" approaches where research priorities determined at the experiment station level are transmitted down to farmers, who are not directly consulted in the research process.
5. There is of course no assurance that the farming systems approach to research will always give results that meet expectations. However we believe that application of this approach can help give a more realistic evaluation of what is possible at the farmers' level and can help address the needs of farmers with different characteristics.
6. In the long run, agriculture will, of course, decline relative to the industrial sector in terms of its contribution to GDP, exports, and employment. However, the proven small size, limited rates of growth, and/or labor absorptive capacities of the industrial sectors in many developing countries means the majority of the populations will for a long time to come continue to derive their incomes from agriculture.
7. However this, as we discuss later (Chapter 6), often depends on the population density. Traditional farming systems using shifting cultivation techniques certainly had this characteristic.
8. Agronomy, Plant Science, Animal Science, Crop Protection, Soil Science, and Agricultural Economics and Rural Sociology.
9. Some academics resent the idea of tied research funding as an infringement on academic freedom. However, we believe that the developing world cannot afford the luxury to finance work not relevant to development problems. Whenever possible, encouragement should be given to using the intellectual talent and available financial resources for work on priority research problems. It is

unfortunate that in some academic circles, such talents are not fully used due to lack of finances for supporting research. It is to the advantage of IAR that it is a part of ABU, which permits the split appointments. In advocating a tied research funding strategy we recognize that approved research projects are likely to be applied in nature and often have a short-run focus. Basic research projects, especially those with a long-run orientation, are less likely to be approved. We do not wish to underestimate the potential value of the latter type of research. However, funding and staffing limitations within national programs and often the necessity of critical masses of research support (both personnel and equipment) has led us to the conviction that such projects are usually best undertaken outside the mandate of national research programs.

10. This now resembles more of an autonomous institution while in addition IAR no longer includes livestock research which has now been incorporated under the National Animal Production Research Institute (NAPRI) (Idachaba 1980).

11. It is in fact still the only agricultural research institute in the country with a substantial socio-economic research input (Idachaba 1980).

12. The department, formed in 1964, previously had been involved primarily in teaching.

13. The type of research relevant to this situation would be very different from that using compulsion and resettlement (Norman and Simmons 1973).

14. This task can be seen from two perspectives, that of the farmers and that of society at large--as articulated by government. It is possible that what farmers perceive as desirable will conflict with government's interests (e.g., subsidized fertilizer distribution) or that government's concerns (with maintenance of long-term soil fertility, for example, through improved conservation measures) may be at odds with farmers' shorter run interests of surviving until next year.

15. Assessment of such programs could perhaps be more usefully done by planning units in government but at the time these were poorly developed in the northern states.

2
The Farming Systems Approach to Research

"Aid that works requires human contact between the helper and the helped. There has to be that vital communication. Go to the villages! Talk to the people. Find out their problems and needs . . . and make sure they are involved."
Critchfield (1979)

In the next four chapters, we describe the environment in which savanna farmers in West Africa operate. It influences in two ways what any rural household can and cannot do. The technical elements of the environment--rainfall, temperature, soil type--establish certain physical and biological constraints on agricultural production systems. Sorghum and mangoes are possible; rye and peaches are not. The socio-cultural or human environment in which any household lives also limits its behavior and that of individuals within it. Community norms and beliefs exercise considerable immediate control on life at village and household levels. National institutions and objectives exert a pervasive influence on the social and economic structures that evolve as modernization and development occur as well as specifically affecting certain factors--taxes, prices for export crops, money supply, and the like. Finally, what farming households do today is influenced also by what happened in the past. Their relationships with and actions in their present environments are conditioned by historical or traditional knowledge as well as by applications of new or innovative information.

In this chapter, a conceptual model of a farming system, which takes these various elements and factors into account, is developed. The model enables the analyst to adopt a holistic view of the farming environment rather than a narrow--and perhaps more usual--commodity or resource view. Recognizing the pivotal role farming households play in determining actual farming systems, we outline a farming systems approach to research. This research approach involves farmers and their households directly in the process of agricultural research. Their involvement in farming systems research, it is posited, will increase the efficacy of agricultural research by helping to ensure that relevant improved agricultural technology is developed and adopted by farming households. Further, the farming systems perspective that emerges from farming systems research applications is useful in highlighting

15

critical issues in agricultural development and in designing more appropriate development strategies and support systems. The conceptual framework of this chapter underlies much of the empirical material upon which the rest of this book is based.

THE ENVIRONMENT OF THE FARMING HOUSEHOLD

A system can be defined conceptually as any set of elements or components that are interrelated and interact among themselves. Thus, a farming system results from a complex interaction of interdependent components that bear upon the agricultural enterprise of a rural household. At the center of the interaction are farmers themselves, exercising some measure of control and choice regarding the types and results of the interactions. To farmers, the means of livelihood and the social and cultural welfare of their households are intimately linked and cannot be separated. We will, then, frequently refer to the farming household or family rather than just the farmer.[1] The members of the farming household, in achieving a specific farming system, allocate certain quantities and qualities of basic types of inputs--land, labor, capital, and management--to three processes--crops, livestock, and off-farm enterprises--in a manner which, given their knowledge, will maximize goals they strive to reach.

Figure 2.1 illustrates graphically some of the possible underlying determinants of a farming system. The "total" environment, as we term it, in which farming households operate can be analytically divided into two parts: the technical (natural) element and the human element (Norman 1976).

The natural resource endowment, or technical element, in any given location restricts what the farming system can be; it, therefore, provides the necessary condition for its presence. In agricultural research, as usually defined, the technical element receives the most attention, particularly, as might be expected, from the technical scientists. They seek to enhance water availability through irrigation and soil quality by applying fertilizer. Methods of tillage and erosion control as well as regimes of micro-nutrient supplementation and herbicide application are similarly tested to overcome the physical deficiencies of the resource base. Manipulating the biological factors forms a separate but related area of technical concern. Scientific inquiry into crop and animal physiology, disease, insect behavior, etc. enables fundamental changes to be incorporated into the organism itself. Plant breeders, for example, alter the genetic structure of particular plants to emphasize the desired characteristics (such as yield, length of stem, insect and disease resistance) while eliminating others (such as drought proneness, off-color of grain, etc.).

Technical scientists have had considerable success in developing crop, livestock, and agronomic systems that can modify the technical environment and improve the potential output of a farming system. Any actual farming system, however, is a subset of what is potentially possible in technical terms. It is the human element that provides the sufficient condition for development and utilization of a particular farming system.[2] People decision-makers

17

FIGURE 2.1
Schematic Representation of Some Farming System Determinants

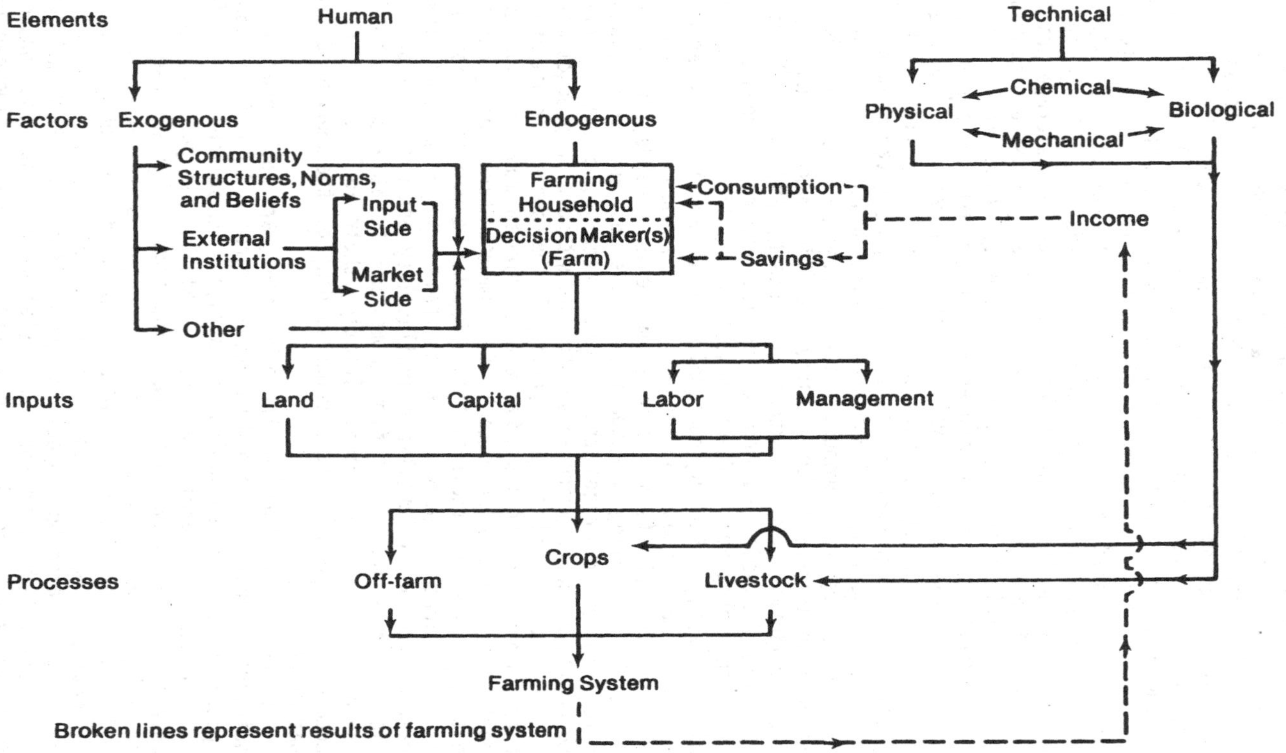

take into account two sets of social, cultural, and economic factors: those exogenous and those endogenous to the household.

Exogenous factors are largely outside an individual's or a farming household's control but they influence what its members will do and/or are able to do. Two aspects of the social environment[3] account for significant variation among farming households.

First, community structures, norms, and beliefs affect virtually every decision of resident households. A long-established household at the top of an authoritarian local political structure is not likely to behave the same way as a newly-arrived stranger-household or as a freed-slave household. On the other hand, both new and established households are likely to share certain food beliefs that affect crop choices and storage practices.

Second, external institutions, especially those associated with central political or governmental authority, also exercise a degree of influence on rural household and community behavior. The strength of this influence varies with the location of the household and its accessibility to outsiders, and with the degree of market involvement. From a farmer's point of view, government activities in supplying agricultural inputs and information and in controlling or stimulating the product markets may be the most important external influences on agricultural decision-making. On the input side, programs such as extension, credit, and seed distribution often are financed and managed by the central government. Such government programs reflect government policies and priorities fairly directly. Farming households may or may not have access to, or wish to have access to, these external resources. The government may also directly (e.g., through marketing boards) or indirectly (e.g., by improving transportation systems and crop evacuation routes) influence prices that farmers receive, and thus influence the choice of crops produced and the amount sold in product markets.

Both these and other aspects of the social environments in which farming households live are liable to modification and, of course, to change over time.[4] Governments take policy decisions that may have as their initial impetus a deterioration in the national balance of payments and as their final farm-level impact, a drastic cutback in fertilizer availability and rise in staple grain prices. While individual farmers may not be able to control such changes in their environment directly, their ability to survive or succeed will depend on their understanding of the exogenous factors that affect them.

Endogenous factors, on the other hand, are those over which the individual farming household has some control. Allocating labor and capital as well as developing and applying management skills follow from internal household decisions. Still, the partial nature of control must be emphasized here. The acquisition and use of land, for example, is by and large an endogenous variable under most conditions. One farmer can clear a new field from the bush or plant his entire farm in cotton, if he decides to do so. But such decisions may also be subject to the exogenously determined overall availability of land or the status of the user-household. Similarly, the household's use of capital may be influenced by the exogenously determined availability of credit and to management skills tied to supplemental extension inputs exogenously provided.

A farming system obviously is complex, which is the reason that improved agricultural technology--thought to be well suited to a particular agroecological situation--has often not been adopted, or why the degree of adoption has varied so widely among farming households. The farmer as decision-maker has until recently received little attention in agricultural research in developing countries, particularly in Africa and Latin America.[5] But it is now increasingly recognized that, without considering the human element as well as the technical element, agricultural research often will not result in relevant agricultural technologies--and the expected benefits may not materialize.

In the next section, we consider the farming household's decision-making objectives, and posit that agricultural research results will be relevant to farmers' own farming systems only when the farmers perceive that the results can enable them to achieve their goals.

GOALS AND THE PIVOTAL ROLE OF THE FARMING HOUSEHOLD

In conventional economic analysis, it is usual to assume that the motivating force of people is utility maximization, that is, getting the most of some value--pleasure, income, food, or goods and services, for example. In agricultural households, the desire to attain maximum welfare results in setting a goal or goals to govern farming decisions. Objectives affect both the way resources are used and the level and combination of processes which are undertaken, such as one off-farm job and five crop enterprises. The underlying goals are also important in determining the degree to which farming households may be willing, or indeed be able, to obtain extra resources to commit to operating their farming operations. A goal of household food self-sufficiency, to cite one possible goal, would imply that emphasis would be firmly placed on marshalling all household resources (land, labor, etc.) to guarantee food supply. In such a case, the potential for entering the market place to obtain extra resources may be extremely limited.

One of the basic tasks for developing and assessing the potential for adoption of improved technology, therefore, is to evaluate its compatibility with the goals as well as the resources (land, labor, capital, and management) of farming households. But determining what the goals are and carrying out the evaluation requires that the analyst be closely in touch with the farming household and the technical and social environment within which it operates.

Three factors seem particularly important to this task: the time frame of farmers, their multiple goal structures, and the conventional analyses' lack of suitability.

The Time Frame

The time frame to which the household decisions are keyed and to which the goals are applied may play a significant role in a household adopting or rejecting improved agricultural technology. While much has been made of the nonchanging, age-old traditions of agriculturalists in the savanna, we have observed considerable change in Nigeria over eleven years. Some modifications in farming operations were in response to the expected climatic variations and demographic change, but other adaptations made by "traditional" farmers were influenced by prices, new seed availability, transport conditions, and other factors.

It is apparent that farming households operate in a dynamic rather than a static situation. This implies that analysis is in a multi- rather than a single-dimension setting--with one critical dimension being time. The farming household today is partly a function of what happened in the past. Historically, in India, for instance, oxen were used as draft animals in the settled farm sector. In savanna West Africa, however, the management of livestock has been in the hands of nomadic herding groups, and settled farmers did not rely on the power of animals for field operations. Only gradually, through commercialization of agriculture, are draft oxen being introduced into West African agriculture.

Similarly, in the situation of extensive availability of cultivable land which has typified the savanna for centuries, there were no compelling reasons to develop sophisticated means to preserve soil fertility. Occasional fallows, animal manure applications, and certain methods of tillage were considered adequate. With growing populations, however, increased intensification and permanently cultivated fields have become the rule rather than the exception. Shortened fallows are one sign of this trend. As Harwood (1979) noted elsewhere, farming households operating near the subsistence level are not likely to be able to forego a portion of their current production potential--as fallowing dictates--for the chance of higher production in the future. Consequently, the need to pursue a short-term private goal of survival could well necessitate sacrifice of a long-term private or societal goal of maintaining soil fertility. Consequently, the chances of farming households succeeding in the long run may be reduced by the actions they need to take to survive in the short term.

Multiple Goals

Farming households in the savanna areas of West Africa are often described as being somewhere on the continuum between subsistence and fully commercialized agricultural production. While complete self-reliance on household resources and production is not the case, neither are household decisions completely tied to market factors. Additional complexities are introduced, moreover, when attention is focussed on the decision-making situation within farming households.

The traditional household farming organization in the northern Nigeria savanna was premised on a group of jointly cultivated fields and a certain number of private fields cultivated in their own time by adults within the household. There appears to have been a significant increase in the number of individual decision-makers within farming households, although the two levels of decision-making with regard to farming operations within the household have not yet disappeared. Even where joint farming households persist young men, and sometimes young women, are often given partial autonomy in control over land and other resources. As a result, the household head, who is responsible for providing food for the household, is likely to have a different goal from other individuals in the household who undertake processes independently. Furthermore, households are still units within specific communities and the exogenously determined rules of social interdependence still apply. Consequently, any farming household may have a complex mixture of goals.

Ancey's (1975) organization of the goal/decision-making interaction which he found in a survey of various parts of West Africa illustrates well one approach to identifying who holds what goals. He specified fourteen different goals and determined at certain levels of decision-making--both within and outside the household--the various goals which were held. The household head, in Ancey's study, emphasized food self-sufficiency, inter-annual security, leisure, prestige, cohesiveness, satisfaction of social consumption objectives, and land tenure prerogatives. The other males in the household emphasized a contrasting set of goals: marketed production, nonagricultural income, net monetary income, and autonomy.

Conventional Analyses' Lack of Suitability

There are, however, severe problems in fitting the multiple goals of such farming households into the marginalist analytical framework conventionally used in investigating fully commercialized agriculture. Such a framework usually implies a single goal of profit (income) maximization. Yet, by itself, this type of analytical procedure is valid only when the welfare of farming households is maximized as a result of pursuing profit maximization, which is rarely true.[6] Market forces do not completely determine welfare, as would be implied by the goal of profit maximization.

The welfare or well-being of farming households in reality appears to consist of two major components: the tangible and intangible. Part of the tangible component, for example, production entering the market place, can be directly measured by market forces. The remaining production, such as that stored in granaries for household consumption, can have a value imputed to it on the basis of market forces--say, the price of buying an equivalent amount at the time of consumption or the price for which it could be sold at the farmgate. That the imputed value truly reflects the welfare obtained for a household placing a high value on a food self-sufficiency goal, however, can be debated.[7] Even more debatable is the ability of the analyst to incorporate in a similar single measure an indicator of household welfare with regard to the intangible component.

As Castle (1977) emphasized, an important part of the welfare of farming households in subsistence societies is manifested in their relationship with the communities in which they live. The social interdependence that exists has both costs, in terms of moral and cash obligations, and benefits, which are realized especially in times of adversity when a localized social security system comes into play. Increased economic independence resulting from rising external economic influences nearly always decreases social interdependence. While shadow prices can provide an analytical way of approaching social costs, it is rare for a factor such as decreased social interdependence to be included when examining welfare benefits (maximized utility) potentially associated with the adoption of an improved technology.

Nevertheless, relationships within the community are important and certainly do influence the goals of farming households and the farming systems they adopt. Harwood (1979) cites an example of a farmer obtaining high returns from a particular improved technology one year but not using it the next year because it had a deleterious effect on his relationships with other members of the community.

DEVELOPING SUITABLE TECHNOLOGY: FARMING SYSTEMS RESEARCH

Farming Systems Research

Farming systems research (FSR), which to some extent resembles farm management as practiced in land grant universities earlier this century (Gilbert, Norman, and Winch 1980), grows from recognizing the interdependence and interrelationships of the technical and human elements within the farming system. As such, it implies a more holistic orientation than is evident in the reductionist approach often used by technical agricultural scientists.[8] The latter approach involves studying one or two factors at a time while attempting to control all others.

Farming systems research also differs significantly from the more common experimental approaches in developing countries in that it involves the farmers themselves, not only as the potential users of the results of the research--as has traditionally been the case--but also as participants in the research process.[9] This means that farmers help to identify the research problems as well as take part in testing possible solutions.

Explicitly recognized is the value that farmers' knowledge, based on their experience and annual experimentation, can play in improving the farming systems they are following. At the same time, their involvement increases the possibility of developing improved systems that will address the constraints they face, be compatible with the goal(s) they have adopted, and, building on the successful parts of the system they already follow, will result in a new response surface which is a combination of the new and old (Harwood and Price 1976).

Including the farmer in the research process also affects the process by opening up new methods for analyzing problems and reaching solutions. The more traditional approach of systems work has been to use models--experiments, linear programming, researcher-managed on-farm trials, etc.--which simply, to various degrees, simulate the real system in laboratory conditions. Farmer participation farming systems research, however, means that the system itself can be incorporated in the experiment and realistic results can be obtained through perturbing the real system via farmer managed trials.

The term farming systems research has often been used loosely. As used here, it has the following characteristics:

1. In the FSR process, the farm or production unit and the rural household or consumption unit--which in the case of farmers in the savanna region of West Africa are often synonomous--are viewed in a comprehensive manner. The whole farm/rural household perspective is included in the research process to focus explicit attention on both the goals of, and constraints in, the farming system.
2. The choice of priorities of research reflects the holistic perspective.
3. In undertaking the research program, research on a subsystem[10] can be considered part of the FSR process by recognizing and accounting for its connections with other subsystems.
4. The results of the research are evaluated not only in terms of the subsystem or subsystems considered but also with respect to the farming system as a whole.[11]

Obviously, the methodological complexities of undertaking FSR can be great because of its systems focus and its holistic characteristic. Therefore, in operational FSR programs, as the above discussion implies, the concept of the total environment--consisting of both the technical and human elements--is preserved. Instead of assuming that all the factors determining the actual farming system can be potential variables subject to manipulation, however, some are treated as parameters not subject to manipulation. For any given FSR program the mixture of variables to parameters is determined by the mandate of the institution in which the program is located, effectiveness of linkages with other institutions (i.e., both of a research and implementation nature), and resources available such as time, skill, manpower, and finance. A limited mandate, few political or communication linkages, and constrained resources are likely to imply a focussed FSR approach with few variables and many parameters. A broad mandate, a high degree of political support, and substantial resources may make an open-ended FSR program with more variables and fewer parameters feasible.[12]

An example of a focussed FSR program would be the work undertaken by the International Maize and Wheat Improvement Center (CIMMYT) in Mexico. This program concentrates on raising the welfare of farming households through introducing/improving the production of corn or wheat with the least harmful interaction with other crops and parts of the system. Another focussed FSR program would be the rice cropping systems work undertaken by the International Rice Research Institute (IRRI) in the Philippines. On the other hand, an example of work more closely approximating an open-ended FSR approach is being undertaken by the Experimental Units of the Senegalese Agricultural Research Institute (ISRA). ISRA attempts to work with both crops and livestock and seeks to have an impact at both farm and policy-making levels.

The primary aim of the FSR process--whether focussed or open ended--is to increase the farming system's productivity in the context of the entire range of private and societal goals, given the constraints and potentials of the farming systems that farming families currently practice. Increased productivity is served by developing relevant improved technologies together with complementary policies that increase the welfare of farming families in ways that are useful and acceptable to them and society as a whole.

The first strategy, which involves developing relevant improved technology, and includes farmer related research, is conventionally known as FSR. The second type of strategy, which to date has received less attention, involves not only farm-level research on improved technologies but also applies the view of the farm to policy issues--that is it links the micro and macro. This strategy is encompassed under what we call the farming systems perspective, which we discuss in greater detail below.

FSR will probably result in recommendations for change that involves small adjustments in farming systems rather than a complete transformation or revolution of technology. Traditionally, for example, many farming systems have exploited the obvious complementary relationships that develop through crop-livestock interaction. Conventional commodity-based research programs have often either discouraged, ignored, or de-emphasized the potential for improving the results of such interaction. As a result, commodity-based recommendations have not been adopted, or when they were, the benefits of the crop-livestock interaction were lost. The FSR approach, as it focusses on evolutionary adjustments in farming systems, should ensure that such complementary relationships and the benefits to be derived from them are not overlooked. Indeed, an FSR approach would be likely to address the increased exploitation of such complementarities, if they were beneficial.

A schematic framework of a farming systems research program is given in Figure 2.2.[13] The research process is recognized as being dynamic and iterative--with both backward and forward linkages between farmers and research workers. Experiment station trials by technical scientists are not eliminated--indeed, they are an integral part of the process--but the linkages between the station and the farmer are no longer one way. The "feed in" from the farmer helps to set station priorities and problems; the "feed back" from the farmers lets the stations' scientists know if they are providing useful results.[14]

Four stages of research can be delineated in an FSR program: stages involving description and diagnosis, design, testing, and extension. The following sections are devoted to a brief overview of each of those four stages with detailed discussion of the methodological and implementation issues postponed to Chapter 7.

The descriptive and diagnostic stage. The objective of this stage is to understand the farming systems that are practiced in the selected target area. This enables the FSR team to determine the constraints that the decision-making households face and the flexibility that exists in the current farming systems--timing, skills, slack resources, etc.[15] The depth and extent of the descriptive work undertaken to achieve this objective may vary, depending upon the treatment which the FSR team conducting the diagnosis can or wishes to make. The more open-ended FSR approach implies that a considerable amount of descriptive information--much of it quantitative--may be needed; a focussed FSR approach will

FIGURE 2.2
Schematic Framework for Farming Systems Research

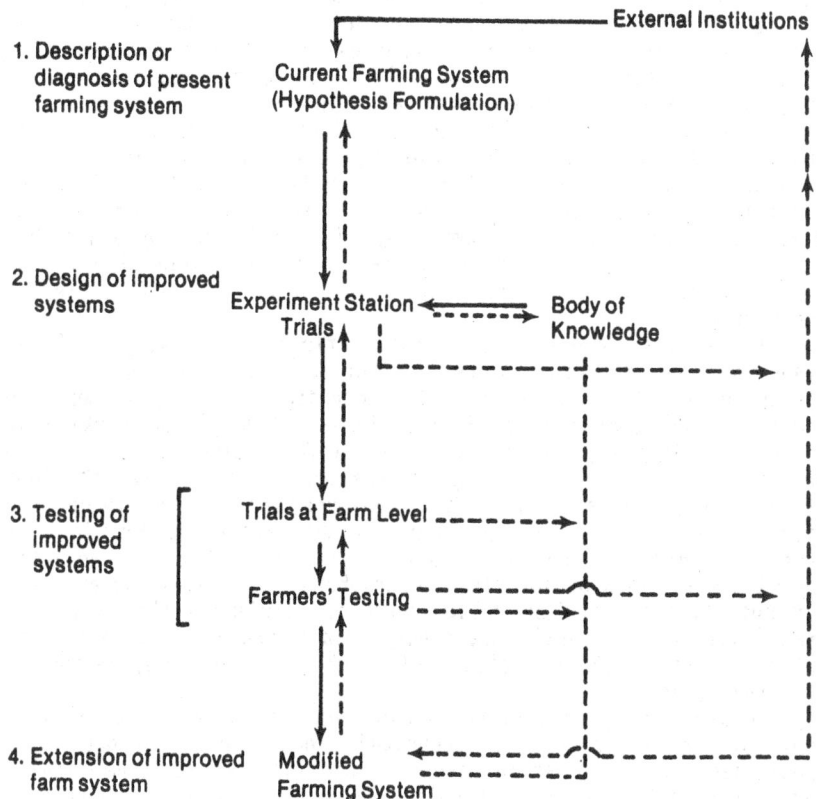

likely require less. In both cases, some idea of the variables
endogenous to household decision-making will be needed to supplement
the more apparent information on technical elements and on exogenous
factors in the socio-economic environment. The goals and
motivations of farmers, which will affect the degree and type of
effort they will be willing to devote to improving the productivity
of their farming systems, are essential inputs to the process of
identifying or designing potentially appropriate improved
technologies.

Obviously it is very likely that within the target area farming
families will be differentiated by both technical and human
elements. The descriptive stage should also yield the basis for
dealing with heterogeneity in the farming population. While in
practice a good amount of judgment goes into identifying homogeneous

subgroups, the objective of population disaggregation can be described scientifically--that is, maximizing the variance between subgroups and minimizing variance within subgroups (Technical Advisory Committee 1978) although this requires more hard data to implement. In other words, the objective of this process is to ensure that farming families within each specific subgroup face roughly the same problems and development alternatives so they should react in the same way to changes in policy and technology. Ideally the descriptive work should classify farming families in subgroups which tend to have similar crop, livestock, and off-farm activities and follow similar social customs, have similar access to support systems, comparable marketing opportunities, and similar technology and resource endowment[16] (Collinson 1981). Although variation in the technical element is easiest to identify and measure, technological improvements may be constrained by other than physical or biological factors; variation in the human element may also be important in addressing the constraints of different subgroups.

There are, however, some complications in carrying out this stage of the research process. For example, recognizing and focussing on the interaction of the technical and human elements in the total environment requires a multidisciplinary team working in an interdisciplinary manner[17] not only at this stage of the research process but also at later stages. The more comprehensive the look at the current farming systems is intended to be, the longer the time needed to do a thorough job and the greater the variety of people needed to do the task. The skills of economists, sociologists, anthropologists, geographers, political scientists, and nutritionists may be called upon to complement the skills of traditional agricultural researchers--plant breeders, agronomists, entomologists, soil scientists, etc.--which, of course, is easier said than done.

Academic or professional disciplines develop as a specialized body of knowledge grows, technical vocabularies become more specialized, and discussion across disciplines becomes semantically and conceptually complex. Farming systems research may require a new breed of agricultural researchers altogether--grounded in one discipline but with more than a passing knowledge of several others. For truly multidisciplinary teams of researchers to work, truly interdisciplinary people are needed, but they will not be trained naturally in current academic systems.[18] So considerable effort will be needed in many cases to develop the capacity to implement a farming systems research approach. Without such effort, however, farming systems descriptions will continue to be heavily biased by the particular personality and disciplinary compositions that thus far have characterized most FSR programs. The technology selected as suitable for developing and testing in stages two and three (described below) will thus be suitable only to those elements included in the description. The chances of their succeeding may be only marginally improved over chances expected if a random lot of available technologies were tested.

In ideal situations, the comprehensiveness of the effort in this stage should be in inverse proportion to the amount of information on farming systems already known. Unfortunately, in many cases, the breadth and depth of the descriptive effort is directly related to the budget supporting the research or the available personnel or to the adequacy or inadequacy of institutional memory. This means that any given FSR program may bear little or no relation to the knowledge already gained or needed. Time and people then determine the task rather than the other way around. The temptation to reinvent the wheel is as strong in farming systems approaches as in any other sort of research.

Several recent applications of FSR, however, indicate that the time required for description can be successfully reduced if: first, the multidisciplinary teams are effectively coordinated for short term, intensive surveys of farmer households; or second, if the areas for detailed inquiry are carefully defined to receive the bulk of attention and other aspects of the farming systems are covered in only a general way; or third, if descriptive work plans are closely linked into the on-farm testing phase--in which case the feedback mechanism provides a broader spectrum of information on conditions and constraints than needed for a more limited test.

The design stage. In this stage, improved technologies thought to be relevant to overcoming or avoiding the constraints identified in the first stage are specified. The body of knowledge (Figure 2.2), the cumulative store of information resulting from other research,[19] is obviously an important source of ideas for potential improved technologies that might be appropriate. In other cases, new technical breakthroughs may have to be sought to address certain constraints; there may be nothing "on the shelf" that will work.

The decision on how to deal with the constraints will depend on the circumstances. Factors to consider include the severity of the constraint, the degree of flexibility in the current farming system, and the availability of potential improved technologies either to break the constraint or to exploit the flexibility in the current farming system. Fine tuning transferred improved technologies to fit local total environments is often possible when constraints are not completely binding and some flexibility for adaptation exists. Where appropriate improved technologies are not available or existing technologies simply don't fit into a constrained rigid environment, the FSR team can in this design stage identify and promote priorities for new research in the programs contributing to the body of knowledge.

The improved technologies, which the FSR team thinks will address the needs of farmers, may sometimes be tested further under experiment station conditions before being sent for testing in the farmers' environment. The need for testing under experimental station conditions will be determined to some extent by the body of knowledge available to draw on or by the degree of risk which the improved technology carries. If the number of variables to be investigated is large or if elaborate insurance schemes would be necessary to prevent farmer loss, experiment station testing as a design task only, makes sense. Normally, however, careful application of suitable information from prior results elsewhere will reduce the need for experiment station work.

The testing stage. The objective of the testing stage is to evaluate a few of the more promising technologies arising from the design stage on farmers' fields. The criteria for evaluating the technologies will be the same as used in the design stage, which, in turn, were those derived in the descriptive stage.

The two parts to this stage are undertaken on farmers' fields. The first, which may not always be necessary, consists of trials at the farmer's level that use farmers' land and labor, but with the managerial input still provided by the research team. The second involves farmers' testing, providing their own land, labor, capital, and management. This on-farm experimentation under farmer-managed trial techniques will provide the potential for substantial research worker/farmer interaction and result in assessing the technologies under conditions as close to reality as possible. That will enable the research team to ascertain realistically the potential suitability of the technologies and possible replicability of results in other similar total environments.

It is in this testing stage that involving extension personnel can particularly strengthen the program. By opening the lines of communication between extension personnel and farmers at this early stage of technology improvement,[20] the contributions of both to further work on the technologies can be elicited and the potential for replication substantially enhanced.

The extension stage. At this stage, technologies found during the design and testing stages to overcome best the constraints delineated in the descriptive and diagnostic stage are widely extended to other farmers. This stage should also be the beginning of the next cycle of farmer feedback and input into the research process. Problems in the extension stage should be monitored--perhaps overlapped with a new round of description and diagnostic work.

Requirements for a Suitable Technology

In general terms, we have already implied that a suitable agricultural technology is a way of doing things (combining resources to carry out processes) that is compatible with environmental constraints (both technical and human) and contributes to the goals and aspirations of the group or individuals using it. The definition of a relevant or suitable improved agricultural technology follows: it is one that is adopted by farming households and improves the efficiency with which they do things. Although that is an intuitively comfortable definition, such a micro- or household-oriented definition may not imply adequate criteria for judging a technology's suitability at the societal level and further specification may be needed. This is discussed later.

The suitability or relevance of improved agricultural technology at the farmers' level has commonly been assessed in an ex post sense, using various methods of acceptance testing, diffusion rates, and the like. Although ultimately such ex post assessments provide the best tests of the suitability of an improved agricultural technology, efficient use of research resources indicates that attempts to assess the potential suitability of

technologies before they are disseminated make sense. Why spend
time, effort, and money devising a better mousetrap if everybody
already owns a cat and is worrying about termites?
Predicting and testing potentially suitable technologies can,
and should, be an integral part of agricultural research. The first
three stages of the farming systems research process explicitly
focus on this task. Conceptualizing a farming system provides a
systematic basis for forming evaluation criteria in terms of both
necessary and sufficient conditions.[21]
The necessary conditions for suitable improved technology,
which determine whether farming families could adopt the technology
if they are willing to, relate to the technical element and
exogenous factors. These conditions can be specified in terms of
evaluative criteria: technical feasibility, community or social
acceptability, and compatibility with external institutions,
especially infrastructural and governmental support systems.
The first criterion has long been accepted by agricultural
researchers as an appropriate one. The various soil, water, and
temperature specifications are part of the regular testing criteria
on most experiment stations. Improved technologies are rarely
released without recommendations as to the physical and biological
conditions needed for the technology to be suitable. The farming
systems research process, however, implies less reliance on
experiment-station established technical feasibility, by providing a
mechanism for determining technical feasibility under farmers'
conditions--where water control options may be more limited and
where soils may be considerably less well-maintained than at
experiment stations.
The relative significance of the community-acceptability and
external-institution-compatibility criteria will depend on the
extent to which potential adopters are market-oriented and the type
of improved technology being contemplated. With increased contacts
outside the village and increased commercialization of agriculture,
for example, it is likely that for an improved technology involving
a cash crop, acceptability in terms of the community norms and
beliefs will be relatively less important, while the presence or
absence of a functioning input supply and output evacuation system
will become extremely important.
Even where a market orientation appears to be relatively
strong, it is often still critical to assess the possible influence
of community norms, especially where patron-client ties remain
strong and where access to land and hired labor is linked to the
social hierarchy. Nevertheless, in almost any case of improved
technology, unless an input distribution system can provide the new
inputs required and there is a market for the product outside of the
household, the improved technology should not be recommended because
the necessary conditions for its adoption simply cannot be met.
Determining evaluation criteria with reference to fulfilling
the sufficient conditions--that is, those that result in the
farmers' decision to adopt--is more difficult and revolve around the
notion of economic feasibility. Substantial variability exists in
the real world. Rural households' resource bases--qualitatively as
well as quantitatively--and the goals of farming households, often
diverge widely. Some kind of weighting system may be needed; some

of the relevant criteria may have to be assessed qualitatively rather than quantitatively.

Certain criteria seem to have wide applicability, however, and may prove useful as evaluative starting points:

1. Because of the prevalent relatively low levels of living in much of savanna agriculture and the desire for at least partial satisfaction of food needs from household production, risk avoidance through ensuring dependability of return is likely to be an important evaluation criterion (Norman and Palmer-Jones 1977).
2. Once food needs have been met, households often follow a course of profit maximization. This criterion is easier to examine by assessing profitability in terms of the most limiting factor--comparing the improved technology with the one it is designed to replace.
3. Another criterion that may be important is minimizing disruption in the total farming system.

If an improved technology involves a whole series of changes in the current farming system, it may be perceived by the decision-maker as unsuitable. On the other hand, a more profitable innovation as dependable as the practice it is designed to replace without deleterious effects on other parts of the farming system--for example, without diminishing the ability to fulfill the goal of self-sufficiency--is likely to be attractive.

Until now we have concentrated on evaluating the improved systems from the perspective of individual farming families. However, it is also important to evaluate its acceptability from a societal viewpoint.[22] For example, if adopting a particular improved technology resulted in degrading the natural resource base, or a more inequitable income distribution, then short run private returns would come at a long run cost to society. Divergence between private and societal interests needs to be avoided. Unfortunately, this is easier said than done. The micro-macro link discussed in the next section is important in trying to ensure this does not occur.

IMPLICATIONS OF THE FARMING SYSTEMS PERSPECTIVE IN THE DEVELOPMENT PROCESS

The farming systems approach to research involves placing people--their capabilities, their goals, their resources--squarely into the process of agricultural research. It focusses on the objective of raising the potential for generating improved agricultural technology that is suited to the people who are to use it, as well as to their fields. The farming systems approach to research permits the concerns of agricultural research to be effectively extended beyond the limits of physical sciences and into the heart of the development process itself, recognizing that constraints to farmers' adoption of improved agricultural technologies are social and economic as well as technical. Cultural and political institutions as well as ecological systems have to bear the stress of technical change. The farming systems approach to research leads to a farming systems perspective that is holistic and integrative and that encompasses all dimensions of a farming household's reality.

In applying the farming systems approach to research, there is a temptation to narrow the focus--to look primarily at factors that the farming household itself can control and to consider the variables, for example, the exogenous factors an individual alone cannot affect, as givens. Commodity prices, for example, are determined by market forces involving many buyers and sellers or by government fiat, so one farmer's voice carries little weight in determining an appropriate price.

But the fundamental recognition of people's roles as decision-makers leads ultimately to the perception that national well-being also depends in a major way upon the outcomes of people's decisions. Depending on the development strategy chosen, national governments' actions to influence the decisions will be more or less direct.

Depending on the level of understanding of farming systems--why they operate the way they do and the kind of transformations they are undergoing--a national government's chosen development strategy may be more or less effective. If it is understood, for example, that farmers are responsive to prices, then prices may be put as high as possible. But if that is wrong and higher prices do not stimulate greater production, both consumers and misunderstood producers will lose. The process of agricultural development will then be slower than anticipated.

Where farmers' decisions to produce or not to produce a particular commodity are constrained by technical or exogenous variables--lack of suitable varieties, lack of information about cultivation techniques, inadequate supplies of fertilizers--national action may be taken to overcome the constraints. The farming systems approach to research can help to identify the most critical constraints and can contribute to giving the agricultural research a high priority to address the technical constraints. Where farmers' decisions are constrained by what have been classified as exogenous factors (such as community beliefs and norms, prices, markets) or endogenous factors (farmers' education, attitudes, etc.), national actions may have to be more indirect.

But will the farming systems perspective--backed by a solid foundation of FSR--automatically suggest what the national actions should be? Insofar as policy parameters have been specified in the descriptive stage as exogenous variables, yes. Policy constraints, if one looks for them, may be clearly perceived. Policies and programs are as amenable to change as water regimes and leaf shapes. The policy changes will involve a different set of professions, however, as determining the range of potentially appropriate policy changes is not a particularly scientific process. Political scientists, economists, administrators, and politicians will probably be involved in specifying possible policy changes and devising ways to test them in an FSR context. Again, we see that multidisciplinary cooperation is highly important.

The problems of testing policy changes also involve factors other than those of testing possible improvements in agricultural technology. Among them scale, multiple objectives, and the necessary political will, seem particularly important.

Scale

A policy generally lays out a fairly broad course of action. Targeting a policy to affect specific types of people with particular characteristics is administratively difficult. Thus, if price is seen as a key constraint to the adoption of a fertilizer recommendation by 10 percent of farmers, for example, it would be difficult for a government to implement a policy of fertilizer price subsidies to that 10 percent only, while ensuring that other farmers pay full price. The range of actions that can readily be affected by policy changes is thus limited in most cases to those in which all farmers, regardless of other characteristics, can share. Roads, power, other infrastructure, prices, interest rates, exchange rates, taxes, and institutions (including those for marketing and credit) are, however, generally of sufficient scale for policy changes that affect their operation to be appropriate and feasible. Such broad policy changes can thus significantly affect both the improvement and adoption of agricultural technology at the farm level.

Such changes, moreover, can be tested through specific pilot programs, incorporated perhaps into other FSR trials, or implemented on their own. Evaluating the effectiveness of such changes may be less easily confined to a single season or a limited group of farmers than evaluating an improved agricultural practice and such evaluation will require different criteria for measuring success.

Multiple Objectives

A factor most likely to cause problems in such an analysis is multiple objectives of policies--some implicit and highly charged politically. Thus, reducing an export tariff on cash crop marketings may release capital for reinvestment in agriculture, for example, but at the same time reduce national revenue and thus, to some extent, diminish the government's power with regard to revenue allocation. The farming systems perspective, of course, is based on the view that improving the technology of agriculture and the welfare of the farming households is the ultimate goal of agricultural research. But that this goal may be only partially shared by the makers of agricultural policy who may be more concerned with the general welfare of the society, must be taken into account in considering policy changes.

Political Will

Ultimately, a certain amount of political will is necessary to implement effectively a farming systems perspective. Even where policy constraints are clearly operative and alternative policies suggested, and perhaps even tested, revising policy often will require legislative or administrative changes, which, in turn, rely upon political leadership or agreement. Where international interests are affected, mustering the national political will may be crucial to policy revision but may still be ineffective. Nevertheless, without such policy changes and the programs to carry them out, in many instances the improved agricultural technology will be suitable to neither the present farming systems nor the people who operate them.

CONCLUSION

In this chapter, we have examined in a conceptual way the complexity underlying the farming systems practiced by farmers. This naturally led to a discussion of the conceptual frameworks which underlie the farming systems approach to research.[23] Farming systems research methods differ in detail but all seek to develop relevant strategies for improving the productivity of farming systems and, hence, the welfare of farming families. We have emphasized that in farming systems research, efforts are made to improve the efficiency of the research process. It does this through: building on the good parts of the farming systems currently practiced; providing a mechanism for exploiting complementary and supplementary relationships between enterprises; complementing the more conventional research programs contributing to the body of knowledge by fine tuning their results to the local situation--sometimes helping prioritize their research goals; and providing the mechanism for a more realistic ex ante evaluation of potential improved strategies. Finally, we have demonstrated the link between technological change and welfare increases at the farm level and the achievement of national development objectives. Policy-makers who adopt a farming systems perspective are, we assert, more likely to achieve an effective match between farmers' and society's goals--and to improve the chances for agricultural development.

NOTES

1. Another reason for the interchangeable use of terms has to do with the frequent existence of several decision-makers in a single household.
2. Some would argue that the order should be reversed with the human element providing the necessary condition and the technical element providing the sufficient condition for the presence of a farming system. However, we prefer the order we have used in the text since the technical element provides less flexible boundaries within which the human element has to be accommodated.
3. We use this term as shorthand for the social, cultural, political, and economic institutions of the environment in which the farming households operate.
4. For example, changes in population density, ease of accessibility to markets, etc.
5. Economists in developing countries have mostly had a macro-orientation, with very few working for any length of time in the micro area--such as farm management.
6. This is particularly the case for farming households in the savanna areas of West Africa. However, many would argue it is even true in highly commercialized agricultural systems such as are found in the United States.
7. Even the assignment of a shadow price is problematic when a

number of households are involved. Each may place a different price
on its ability to produce enough food for themselves, and finding an
appropriate average shadow price would no doubt drive one back to a
retail or wholesale price estimator.

8. The reductionist approach used by technical agricultural
scientists has been increasingly emphasized in recent years. There
appears to have been a more holistic perspective earlier in the
century in United States land grant institutions since farm
management originated in Departments of Agronomy (Gilbert, Norman,
and Winch 1980).

9. Farmers in fact often used to be more directly involved in
the research process in the United States (Johnson 1981), but this
has become less popular in recent years.

10. Subsystem implies a boundary separating the system from its
environment. Two systems may share a common component or
environment or one system may be a subsystem of another. So a farm
system can be broken down into a number of subsystems--for example,
crops, livestock, and off-farm--which may overlap and interact with
each other (Technical Advisory Committee 1978).

11. The farming system reflects the resolution of the conflicts
between the goals of, and the constraints faced by, the farming
household.

12. Winkelmann and Moscardi (1979) term programs with a small
number of variables to parameters as "FSR in the small," and a large
number of variables in relation to parameters as "FSR in the large."
Zandstra (1979b) has also, with different terminology, drawn this
distinction.

13. Semantically, some would argue that the program we are
illustrating, which involves the active participation of the farmer,
is really "downstream" or applied FSR, as opposed to "upstream" or
developmental FSR, which is largely confined to experiment stations
and rarely involves active participation of farmers (Gilbert,
Norman, and Winch 1980). Unless otherwise stated, we use the term
FSR in the spirit of "downstream" FSR.

14. This, of course, is the ideal but it may at times be
difficult to achieve in practice. Traditional farm management
researchers used to try this, but the lack of farmer experience in
problem identification and specification often made the farmer's
participation less than hoped for.

15. The use of the term team implies more than one person. It
is unlikely, though not impossible, for one person to have a
sufficient basic knowledge of the requisite technical and social
science disciplines to undertake the task by herself/himself.

16. Such as soil types, rainfall regimes, income levels, farm
size, etc.

17. Multidisciplinary suggests involvement of a number of
disciplines while interdisciplinary implies the disciplines working
together rather than independently.

18. This is in contrast to the agricultural scientists in the
early days of the land grant university system. Since most had
grown up on farms, they were often good at perceiving farmers'
needs. Thus the animal husbandry researcher--not the animal
scientist in those days--provided a good dose of the
interdisciplinary approach from his own experiences (Sjo 1980).

19. Such as reductionist experiment station based research, "upstream" FSR work, etc.
20. If possible it is desirable that they should be involved earlier, that is, at the descriptive and diagnostic stage.
21. Although we have attempted to break the evaluation criteria into distinct groups, we recognize that they are not always completely exclusive.
22. The word societal is used to imply some aggregation of farming families. It could, for example, mean the community in which farming families are located or the nation as a whole.
23. The model we have presented is based on one drawn up at a meeting in Mali (Institut d'Economie Rurale 1977). However, many other models with the same basic steps have also been developed. See, for example, Beets (1979), Byerlee et al. (1981), Flinn (1978), Hildebrand (1976), ISRA (1977), Moreno and Saunders (1978), and Zandstra (1979b).

3
Agroecology
of the Nigerian Savanna

"Africa is neither a vast reservoir of dormant biotic wealth nor a Cinderella of poverty. It is worth considering certain features of her productive potentiality."

Phillips (1959)

The West African Savanna lies between the humid equatorial high forest and desert ecological areas (Map 3.1(a)). Roughly 4,950,000 square kilometers in extent, it is bounded on the north and south by the 18.9°N and 8.2°N latitudes, respectively, and on the east and west by the 16°W and 30°E longitudes. The southern boundary dips downward slightly toward the east, although the average width of the savanna is about 1,100 kilometers (Kowal and Kassam 1978).

The natural vegetation of the savanna region is dominated by grassland with varying densities of scattered trees or shrubs (Phillips 1959). The climatic characteristics, especially the level and distribution of rainfall, which results in marked rainy and dry seasons of varied duration and intensity, demarcate five bioclimatic or ecological zones within the savanna (Table 3.1). Northern Nigeria has substantial areas in the Southern Guinea, Northern Guinea, and Sudan zones and a small area in the Southern Sahel zone. The Northern Sahel zone is found only in neighboring Niger and in countries lying to the west of Nigeria--Mali and Mauritania (Map 3.1(a)).

The Zaria, Sokoto, and Bauchi villages, with which this book primarily deals, are all located in the Sudan and Northern Guinea zones (Map 3.1(b)). All have a unimodal rainfall distribution and a substantial amount of arable farming. The Southern Guinea ecological zone, which has a bimodal rainfall pattern, was included in the village-study cycle, but the results are not reported in detail here because farming systems in the zone are quite different. Neither do we include the Northern Sahel ecological zone, where transhumance is dominant and very little arable farming is practiced.

The ease with which we have just defined the savanna and the various zones that lie within it belies the difficulties that this environment presents to people living in it and the variability that confronts the agriculturalist trying to recommend optimal technologies to exploit the natural-resource endowment. In this

38

MAP 3.1
The Savanna of West Africa

(a) Mean Annual Precipitation in the West African Savanna

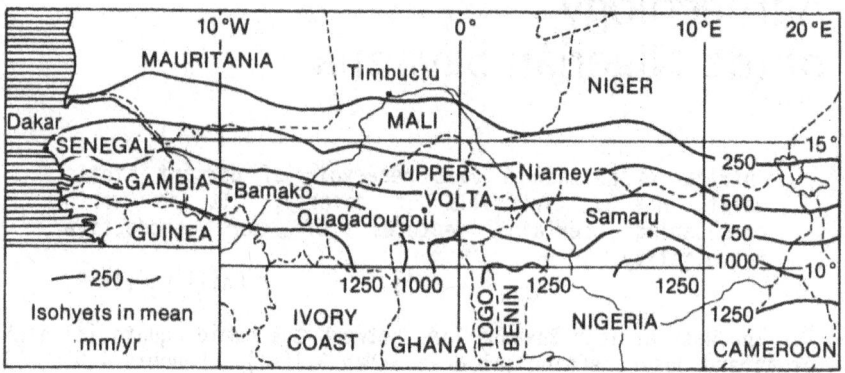

(b) Ecological Zones of the Northern Part of Nigeria

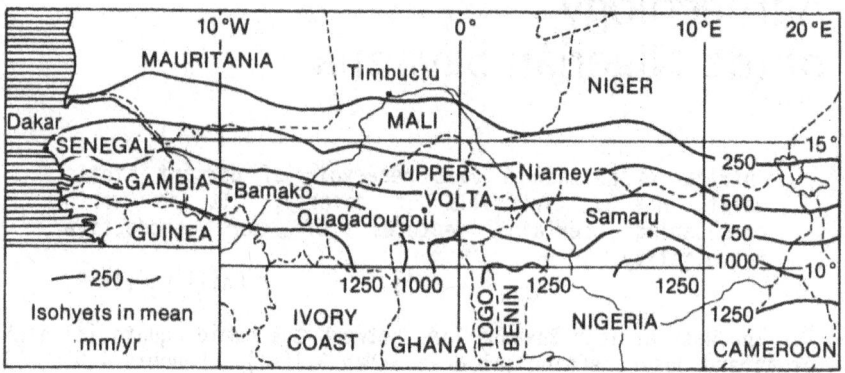

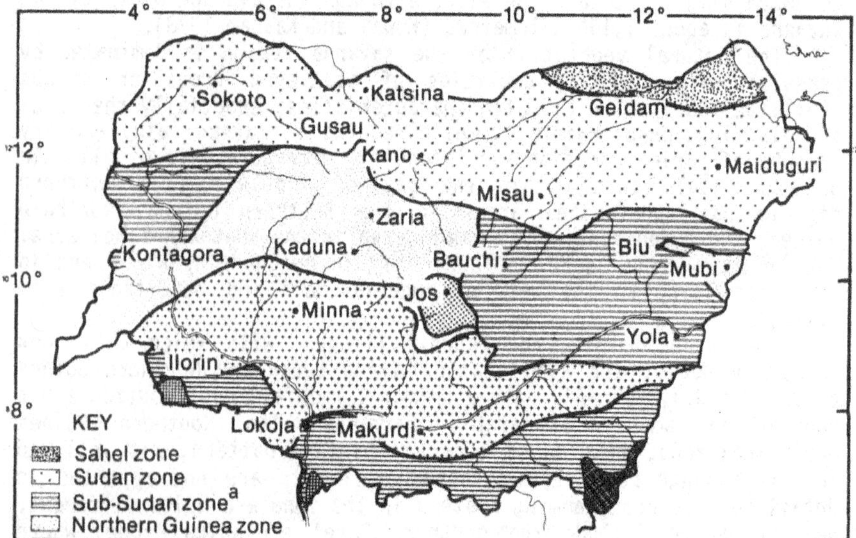

[a] The Sub-Sudan zone is usually considered to be part of the Northern
Guinea ecological zone.

TABLE 3.1
The Ecological Zones of the West African Savanna

Characteristics	Ecological zone				
	Sahel		Sudan	Guinea	
	Northern	Southern		Northern	Southern
Boundary (mm rainfall)	0 to 350	350 to 600/700	500/600 to 380	880 to 1200/1300	1200/1300 to 1500/1600
Length of rainy period (days)	0 to 63	63 to 95/102	95/102 to 140	140 to 187/200	187/200 to 229/244
Main soil types	Sands Arid brown	Arid brown	Non-leached ferruginous	Leached ferruginous	Concretionary ferruginous, ferrisols, ferrallitic
Physiognomy	Open thorn savanna	Open thorn savanna	Shrub woodland	Open savanna woodland	Light forest open woodland
Main tree species	---	Acacia spp. Commiphora spp.	Combretum spp. Acacia spp. Terminalia spp.	Combretum spp. Isoberlinia spp.	Daniellia oliveri
Main grass species	---	Cenchrus spp.	Andropogon gayanus	Andropogon spp. Hyparrhenia spp.	Andropogon tectorum Imperata cylindrica
Main food crops	---	Millet	Millet, sorghum	Sorghum	Yams, maize, sorghum
Main export crops	---	---	Groundnut	Cotton	Soybean, sesame

Source: Kowal and Kassam (1978).

chapter, we describe in some detail the potentials and the constraints the savanna zone agroecology places on the region's farmers. First we discuss the physical characteristics of the technical element; then the biological factors that restrict and affect more profitable use of the savanna ecology; and, finally, the strategies used by both farmers and agricultural scientists to increase production. In addition we comment on the long-term implications of those practices.

PHYSICAL FACTORS

Climate

Climate influences many aspects of crop growth. Radiation and temperature regimes are critical for photosynthesis to take place. But in the savanna areas of West Africa--unlike in more temperate areas where temperature is critical--the most important determinant of crop growth is availability of water. We therefore first look closely at this critical determinant.

Water regime. Kowal and Kassam (1978), in their authoritative reference work on the agroecology of the West African Savanna, have analyzed the rainfall patterns in some detail. For each degree of latitude moving northward from the southern boundary of the zone, rainfall decreases 131 mm. At the same time, rains start 13.4 days later and finish 5.7 days earlier. Because the water regime depends heavily on rainfall, a progressively shorter growing season results (Tables 3.1 and 3.2). That affects the range of crops that can be grown; longer-growing-cycle crops such as sorghum and cotton produced in the south give way to millet-dominated cropping systems farther north. In addition, there is a corresponding increase in the variability of rainfall at the beginning and end of the growing season as one moves north through the savanna region (Cocheme and Franquin 1967; Table 3.2). Thus, farmers in the northern areas often are forced to risk planting at the onset of the first--perhaps unreliable--rains to improve their chances of having a growing season long enough to allow crops to complete their growth cycles. The crucial nature of time that this implies has stimulated much research in the drier Sahel ecological zone on soil preparation and cultivation at the end of the rainy season (Kowal and Kassam 1978).

The water regime does not depend solely on rainfall, however. By considering the interaction between precipitation and evaporation, seasonal variations in the water regime can be divided conveniently into five periods (Figure 3.1), an examination of which reveals the paradox of the extreme conditions of dryness and marginal soil-water reserves at one part of the year and the excessively wet conditions of leaching, waterlogging, and flooding that occur at another part of the year. Therefore, the drainage and management of the soil surface, as well as the control of water and water movement by irrigation, become very important in the savanna region.

The preparatory period indicates the earliest time permitting cultivation (Figure 3.1). Sowing cannot be undertaken, however, until the first intermediate period during which there is a slow

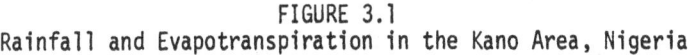

FIGURE 3.1
Rainfall and Evapotranspiration in the Kano Area, Nigeria

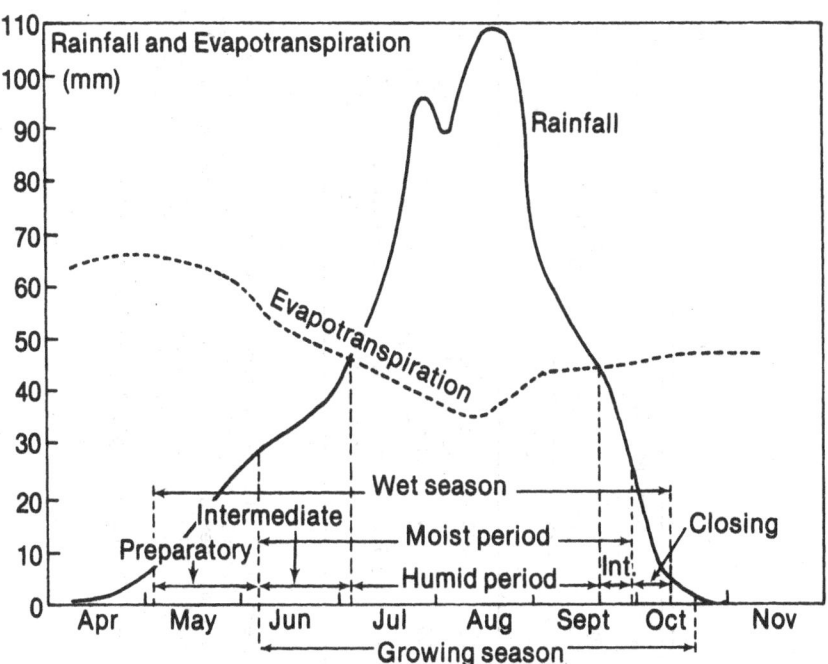

recharge of the entire soil profile with water. There is still some
risk in planting during this period because soil-water reserves are
very low, particularly in the earlier part; and plants will be
largely dependent on rainfall. But during this early part of the
rainy season, mineral nitrogen not lost immediately by leaching is
also released (Kowal and Kassam 1978). Farmers' desire to plant
early, no doubt, partly arises from response to this nitrogen
supply.

During the humid period, too much water can cause problems for
farmers, particularly in the Guinean ecological zones. Increased
runoff and soil erosion are particularly acute during this period
because of the frequent high-intensity rainfall systems.[1] At that
time, additional problems are caused by excessive leaching of soil
nutrients; this increases the difficulties of efficient use of the
highly mobile nitrogen supplied in inorganic fertilizer. Finally,
the high humidity and moisture during this period encourage the
spread and attack of insects and pests.

During the second intermediate period, the drying-up stage in
the annual water cycle, crops mainly depend on the water reserves in
the soil profile. Low levels of soil moisture hasten the elongation
of the root system and the physiological age of the crop (Kowal and

TABLE 3.2
Climate at Three Locations in Northern Nigeria[a]

Area	Ecological Zone	Mean Monthly Temperature		Total Rain (mm)	Length of Rainy Period (Days)	Growing Season			Months When Water is Surplus	Rain for Individual Months (mm)		
		Min.	Max.			Length (Days)	Date Start	End		May	Aug.	Oct.
Sokoto	Sudan	15.0	40.0	752 (18)	120	150	June 1-10	Oct. 21-30	July-Sept.	42 (148)	250 (59)	23 (223)
Zaria	Northern Guinea	13.9	35.0	1115 (15)	150	180	May 11-20	Nov. 1-10	June-Sept.	132 (80)	231 (56)	36 (193)
Bauchi	Northern Guinea[b]	12.8	36.7	1102 (19)	140	170	May 21-20	Nov. 1-10	June-Sept.	91 (90)	346 (46)	37 (164)

Source: Compiled from Kowal and Knabe (1972).

[a]The three locations represent the areas in which the empirical studies referred to frequently in Chapters 5 and 6 were undertaken. The figures in parentheses denote coefficients of variation. The start of the rains and the start of the growing season is defined as the first ten-day period in which the amount of rainfall is at least equal to one-half the evapotranspiration demand. The end of the growing season is assumed to occur when the water storage in the top 10 mms of soil is used up. Water-surplus months are defined as those in which rain exceeds evapotranspiration and soil water storage.

[b]Sometimes the Bauchi area is split off from the Northern Guinea ecological zone and classified into a Sub-Sudan zone (Map 3.1(b)).

Kassam 1978). Physiological maturity is preceded by a loss of dry matter--commonly called senescence.

During the dry period, crops cannot be grown on upland fields without irrigation, although in the southern parts of the savanna, low-lying bottom lands may lie close enough to the watertable to permit production, in that crops can draw on the subsurface water. Unless upland soils are sandy or tractors are used, however, the land becomes too hard to cultivate before the rains start. Nonetheless, one traditional advantage of the dry period can be cited: it provides an effective means of controlling epidemics of pests and diseases.[2]

Temperatures. High radiation and high daytime temperatures which favor high rates of photosynthesis characterize savanna climates. Combined with low night-time temperatures, which decrease respiration losses, such factors contribute to the high rates of dry-matter production and yields of certain crops in the savanna.[3] Since temperatures tend to be 10°C to 20°C warmer throughout the savanna than in temperate zones,[4] so chemical reactions, for example, are two to four times faster. Mineralization of soil organic matter and decomposition of crop residues can thus occur quickly if moisture conditions are favorable. When biological processes are rapid, crops tend to grow faster as long as water and nutrients are not limiting.

Although average temperatures at a given place and time of the year vary relatively little (Cocheme and Franquin 1967), the annual temperature does increase from south to north. That corresponds with an increase in radiation reaching the crop and a decrease in annual rainfall toward the north. The lowest temperatures occur in December and January and can result in delayed growth for irrigated crops such as rice, sugarcane, and cotton. Delayed growth for certain other crops results not from the cold itself, but rather from the wide diurnal variations in temperatures.[5] Germination for most crops is most favorable in the rainy season, when the temperature regime and the radiation characteristics are also most favorable.[6]

Soils

Soils, of course, cannot be measured and averaged in the same way as water and temperature regimes, but they can be usefully classified in various ways. Farmers in the Nigerian savanna generally differentiate between upland, or gona, soils and those located in valley bottoms, or fadama. Ferruginous tropical soils, according to the d'Hoore classification system, comprise the gona in the Northern Guinea and Sudan ecological zones, whereas the brown and reddish-brown soils of the semi-arid and arid areas dominate the upland fields of the Southern Sahel ecological zone (Table 3.3). In all three zones are found extensive areas of hydromorphic soils in low-lying fadama fields. These areas, widely used during the dry season, can be economically important.

Under natural conditions in the savanna, the soil surface tends to be porous and fairly well structured, particularly if it has had prolonged protection against fire. This structure is a result of a high organic content, protection of aggregates against rain splash because of the natural vegetation, and the very high biological activity of earthworms and termites. As a result of fire and cultivation, however, the natural structure of the surface soil is rapidly destroyed--due to a reduction in the biological activity of the soil, a decrease in soil organic matter, and increases in rain splash and soil erosion.

TABLE 3.3
The Relative Distribution of Soil Types in the Savanna of West Africa

Soil type	Area (%)
Ferruginous tropical soils and associated soils	60
Ferrallitic soils	10
Ferrisols	7
Brown and reddish-brown soils of arid and semi-arid areas	5
Vertisols	2
Other soils (mainly hydromorphic)	7
Rock, debris, ferruginous crusts	9

Source: Kowal and Kassam (1978).

With increasing population pressure land has become more intensively cultivated. Increasingly, then, land is more completely cleared and fallow periods are reduced and eventually eliminated. Large, continuous areas of cultivated land raise the potential for runoff and soil erosion, which could contribute to land degradation and the further reduction of soil productivity in the future. Controlling soil erosion and maintaining a productive soil structure become increasingly significant issues. We look briefly here at some of the chemical and physical characteristics of soil that must be taken into account by savanna agriculturists.

The highly weathered, ferruginous tropical soils receiving between 500 and 1200 mm of annual rainfall tend to be very lateritic because of a loss of silica. The soils, usually formed on parent material rich in quartz, tend to have fairly shallow profiles, less than 150 cms deep. A typical cultivated soil profile has a sandy surface and a compact clayey subsoil.[7] Because of those characteristics and, typically, low levels of organic matter, the cation exchange capacity[8] of the soil tends to be low, which reduces the buffering capacity[9] and results in low nutrient cation levels, particularly with respect to phosphate. In addition, free iron oxides tend to be deposited in the profile in the form of mottles, concretions, or even a hard pan. The water-holding capacity of the soils can be reasonable, although that depends on the soils' structural condition. Because most of the soil aggregates are very small and unstable, they tend to compact when wet and to form surface crusts that erode readily. Both nutrient deficiencies and structures of these soils thus create management problems for

farmers.
Phosphate deficiency is a serious problem in most parts of the savanna. Phosphate in commercial fertilizers is commonly immobilized through acidity, although it has also been found that over several seasons the residual effects of phosphate can be beneficial even at fairly modest application levels. That has been the rationale for investigating the potential of rock-phosphate applications, particularly in the francophone regions of the savanna. Further, it has been noted that calcium and sulfur contents are often as important as the phosphate content in determining the shape of the yield-response curves (Kowal and Kassam 1978).[10]

Besides phosphate, available nitrogen, another important crop nutrient, is generally low in savanna soils. Both cereal crops and cotton respond well to nitrogen fertilizers. In most soils, nitrate accumulation peaks early in the rains and is lost later by leaching, unless the nitrogen already has been taken up by plant roots. Early planting is therefore undertaken, as already mentioned, to capture as much as possible of the soil nitrates before they are leached. Because under such conditions the timing of nitrogen-fertilizer applications is critical, split applications often have been advocated, although that does complicate the seasonal work profile for farming families.

Like nitrogen, sulfur in the soil can be lost through burning and leaching. The atmosphere contributes a small amount of sulfur to the soil, but that gain can be outweighed by the losses. Sulfate, however, is conserved by adsorption on the clay of textural B horizons (Kowal and Kassam 1978). Sulfur deficiency commonly occurs in both groundnuts and cotton and, in fact, would probably be commoner than it apparently is were it not that phosphate fertilizers applied generally contain sulfur.

Sufficient quantities of potassium, on the other hand, are usually present in the soils of the savanna, and crop responses to applied potash are rare except under intensive, continuous production on soils formed on noncrystalline parent materials--such as those in fadama areas. Calcium and magnesium deficiencies are also rare but can be brought about by long-term cropping on poor soils. Low levels of calcium can contribute to low shelling rates in groundnuts. Boron deficiencies can reduce yields in cotton, and molybdenum deficiencies sometimes have a similar effect on groundnuts.

In general, because soluble nutrients resulting from weathering or mineralization are rapidly removed or leached during the rains and because nitrogen, sulfur, and other elements are lost through bush fires or burning of crop residues--both of which are common practices--the total quantity of available plant nutrients and bases in savanna soils is small. In addition, the amounts of organic nitrogen, phosphorus, and sulfur mineralized annually are often well below the amounts needed to sustain high crop yields. Lengthy fallows are, therefore, necessary if good yields are to be obtained without adding manures or inorganic fertilizers. Given the population pressure on land in many areas, such fallows are unlikely, so nutrient supplements are essential to maintain reasonable yields.

The brown and reddish-brown soils of the semi-arid and arid areas have parent material that is commonly aeolian in origin; they are found in areas where rainfall rarely exceeds 500 mm. Weathering and leaching tend to be slight and, although the soils' physical properties are reasonably good, the structure tends to deteriorate rapidly when the soils are cultivated. Organic matter is low in clay in the top soil, but cation exchange capacities are reasonable. The agricultural potential of such soils is, however, limited by lack of moisture, making them most suitable for extensive grazing.

BIOLOGICAL FACTORS

Biological factors relate to crop and animal physiology, diseases, and pests. In recognizing the crucial linkage between physical and biological factors, farmers traditionally have manipulated the crop and livestock enterprises to ensure a degree of compatibility between the two types of factors. For centuries farmers have adapted their farming systems (e.g., agronomic cropping patterns in relation to plant stresses), to minimize the adverse effects of constraints imposed by physical factors, by exploiting the biological characteristics of the various crop and livestock enterprises. These enterprises also have various constraints of a biological nature, which we now discuss.

Growth Cycles of Crops

The hydrological phases discussed earlier are critically important in understanding the growth cycles of the crops grown in the savanna. Crops such as early millet (gero), maize, cowpeas, and groundnuts are nonphotoperiodic[11] and fit reasonably well into the available moist period (Figure 3.1). Other local crops--cotton, local sorghum, and late millet (maiwa)--have a potential growth cycle of 160 to 180 days, which is longer than the average moist period. Though it appears that soil-moisture stress induces or accelerates senescence or maturity for these crops, they do have to rely on residual moisture in the soil to fill their grains. Therefore, yields drop quickly if rains come later or end earlier than usual. In addition, these crops are photoperiodic.

Because there is apparently a definite relationship between the amount of rainfall and its variation at the beginning and the end of rains, hydrological factors directly affect the relative degree of emphasis the farmers place on photoperiodic and nonphotoperiodic crops and the routine incorporation of both types into cropping patterns.

Pests and Diseases of Crops

In general, except for cowpeas and cotton, pest and disease problems for the major crops grown in the savanna are not severe (Kassam et al. 1976). Nevertheless, factors of a biological nature at present do inhibit increases in the potential yields of the major crops. The smut diseases (Sphacelotheca sp.) produce the greatest damage to sorghum, although seed treatment is effective in controlling the major types. The most serious disease of millet is green ear (<u>Sclerospora graminicola</u>), a downy mildew. The extent to which it is a problem depends on the variety of millet and environmental conditions. Local cultivars of sorghum generally have adequate resistance to the foliar diseases, such as the downy mildew (<u>Sclerospora sorghi</u>), and that should facilitate genetic resistance in improved cultivars of this crop.

Head mold, caused by a complex of organisms, is commonest on sorghum maturing during the humid period. That factor probably encourages farmers' preferences for photoperiodic long-season sorghums, which head at the end of the humid period or the second intermediate period. The problem of head mold has inhibited the introduction of short-season, improved cultivars of sorghum that have been developed. Early millet is resistant to head mold, but late millet shows less resistance.

Improved cultivars or changes in management practices may introduce other pest and disease problems. For example, it has been observed that under traditional systems of early sowing with local long-season cultivars, the damage to sorghum by midge (<u>Contarina sorghicola</u>), shoot-fly (<u>Atherigona soccata</u>), and the stem-borers (<u>Busseola fusea</u> and <u>Sesamia</u> sp.) and to millet by midge (<u>Geromyia penniseti</u>) is relatively unimportant. However, late sowing and early harvest of short-season cultivars sown in areas where long-season cultivars are also grown could change the dynamics of these insect populations, particularly because they permit pest population build-up from one crop to the next. Consequently, control measures could become necessary and economically significant, either through chemical control or through integrated pest management or genetic control via resistant cultivars.

A major problem for sorghum, millet, and improved cultivars of maize is the damage caused by a semi-parasitic weed called striga (<u>S. hermontheca</u> and <u>S. senegalensis</u>). Heavy emergence of striga occurs toward the end of the moist period when local late sorghum and late millet crops are heading. Striga has particularly severe effects on maize, even before the semi-parasite emerges above the surface. Several methods have been developed for reducing if not eliminating the problems of striga: hand weeding, rotations, high-soil fertility, host resistance, and foliar and soil-active herbicides (Ogborn 1974; King 1972). None to date, however, have proven to be feasible, practical, and economical.

Another problem for millet and sorghum for which there appears to be no effective and economic control at present is the weaver bird (<u>Quelea quelea</u>), which eats grain, except for red sorghum (Crook and Ward 1968).

The major problem in groundnut production is fungal infection of shells and kernels. Aflatoxin caused by <u>Aspergillus flavus</u> makes infected kernels toxic and unfit for human or animal consumption (McDonald 1969). Seed dressing can substantially reduce infection rates and rapid post-harvest drying can reduce aflatoxin development. The major foliar disease of groundnuts is leaf spot (<u>Cercospora</u> sp.). At present no completely resistant cultivars are available, but the disease organism can be controlled effectively by spraying (McDonald 1973). Rosette virus disease, the vector of which is <u>Aphis craccivora</u>, is a major problem particularly in the southern part of the savanna. High seed rates and early sowing have often been recommended to reduce attack.

The most serious pest of the growing cowpea crop is Maruca testulalis, which attacks flowers and pods, causing them to shed (Raheja 1974). Beetles (such as Callosobruchus masculatus and Bruchidius atrolineatus) are also major factors contributing to the storage loss of cowpeas (Caswell 1968); damage in storage can be reduced through minimizing delays in harvesting and storing in the shell. Cotton is attacked by bollworms (such as Diparopsis castena), cotton stainers (Dysdercus sp.), and sucking bugs (Empoasca facialis), which together can greatly reduce both yield and quality. Seed dressing can control bacterial blight (Xanthomonas malvacearum) in cotton quite effectively.

Nutrition and Disease in Livestock

The production of livestock, as well as crops, typifies agriculture in the savanna. Whereas the small ruminants--goats, sheep--and poultry tend to be associated with settled villages, the cattle herd is predominantly transhumant or nomadic. Thus, cattle movements and productivity are related to the hydrological cycles and to other climatic and soil factors already discussed. An annual migratory cycle is practiced. Cattle and their herders move south at the beginning of the dry season, concomitant with the recession of the rains, and move north with the rains at the beginning of the following season. Such a system has many advantages including the following:

1. The annual movement is compatible with the seasonal fluctuation in the quality of herbage and the availability of surface water.
2. The seasonal movement is also compatible with minimization of certain diseases. For example, the adverse effects of the tsetse fly, which advances and retreats with the rains and carries the protozoan disease called trypanosomiasis, are minimized.
3. The move northward during the rainy season permits the use of land that is suitable only for grazing and that will support livestock only during that time of year.
4. The movement south in the dry season permits the establishment of complementary relationships with settled cultivators, in which crop residues provide sustenance for the livestock and livestock provide manure for the fields.

The livestock systems currently practiced involve using grassland low both in terms of quality and in productivity per unit area per year. Grazing lands characteristically consist of grass sparsely distributed over the land, forming unstable vegetation associations. Coarse grasses tend to dominate; these are palatable and nutritious only when young. Legumes tend to be scarce. The growth of vegetation stops immediately with the end of the rains, and the grass dries out rapidly to produce poor-quality hay. Woody species tend to invade the grass (Kowal and Kassam 1978), and when fire is excluded from the area, the invasion is accelerated. So burning pastures early has been traditionally practiced in much of the region, and in addition to producing a flush of green herbage in most areas, the burning encourages the growth of more palatable and productive species such as Andropogon gayanus.

The limited quality and productivity of the grassland inevitably means that livestock productivity is low, in terms of low milk and meat yields per animal unit and per unit area. Though cattle species in the herds are selected for their adaptation to variable conditions, the white Zebu being the most common, the long dry season--linked with exposure and malnutrition during that period--prevents steady growth and normal maturation. In cattle, for example, maturity is delayed until the age of 5 or 6 years and reproductive rates are very low. Animals, because of their relatively poor nutrition, also tend to be more susceptible to the vectors of various diseases such as tsetse flies and ticks.

IMPLICATIONS OF THE SAVANNA RESOURCE BASE FOR PRODUCTION

Photosynthesis of plants--combined with the physical factors, soil and climate--is the starting point of the process of providing sustenance for man and animals. In agriculture, man uses knowledge, skill, and labor to manipulate the aerial, edaphic, and biotic elements of the natural environment to provide food and other materials important for human welfare. The explicit intervention of people therefore has a critical influence on an area's ecology. That influence, depending on circumstances, can be beneficial or detrimental, and it is thus impossible to treat the technical and human elements completely independently of each other. In this section, therefore, the interactive process will be more firmly emphasized than in the earlier discussions. First, we look at some of the farmers' strategies for manipulating the environment to increase production. We then summarize some of the agricultural scientists' suggestions for maximizing resource productivity, concluding with some thoughts on the future of the savanna as a resource base for agricultural growth.

Farmers' Strategies

Over the generations, farmers have adapted their farming systems so as to minimize the adverse effects of constraints imposed by physical factors. Generally they have exploited the biological characteristics of the various crop and livestock enterprises through a variety of farm-management practices. Seed selection and establishing a crop calendar, of course, are important, but we focus here on three management practices that reflect the management of several climatic, soil, and biological factors simultaneously: the ridge and ring systems of cultivation, and the practice of intercropping or mixed cropping.

The ridge system of cultivation. Under traditional systems of farming, tillage is done by hand. As a result, clearing is rarely complete and land is cleared by cutting and burning in situ. In the ecological zones of central interest here, however, some land is often completely cleared, and is cultivated for a long period of time, sometimes even permanently. This permanent cultivation system involves a fairly complex method of ridge cultivation not completely related to climate, soil, topography, or land conservation; rather, it contributes to management of all factors at once. Kowal and Stockinger (1973) have suggested that, in agronomic terms, the ridge system of cultivation presents a number of advantages:

1. It cuts down on the time involved in seed-bed preparation because only half the area is worked. This is important when timing at the beginning of the rains is critical.
2. In the process of ridge preparation, the topsoil, enriched with ash and plant residues, is concentrated in the area of the plant roots. This effectively increases the thickness of the topsoil, thereby enhancing soil fertility.
3. Ridges can protect against soil erosion when used on the contours of the slopes.
4. During the moist period, the ridges improve aeration for the roots of crops planted on top of the ridges, while the furrows can act as open drains. In areas where water is deficient--and rapid drainage is not desired--ridges are often crosstied to conserve both water and soil for crops involving underground parts, such as groundnuts and tubers. The softer, more friable soil is located on the ridge, where crops can be more readily pulled or dug with less loss.

The ring system of cultivation. The ring system of cultivation also reflects the importance of hand labor, and farmers' recognition of the need to supplement soil fertility with additional nutrients. As few farmers have access to significant quantities of inorganic fertilizers or to mechanized transport between homes and fields, the major sources of additional nutrients are animal and household residues and the major power sources to get them on the fields are baskets carried on the head or panniers on donkeys.

The ring-cultivation system implies the existence of a set of concentric rings around the village settlement. Farmers tend to use the upland fields close to their compounds and to the village permanently, maintaining soil fertility in this inner ring through the regular incorporation of organic residues, including household waste. Sheep, goats, and chickens are usually kept inside the housing compound during the beginning of the rainy, planting season, so their droppings are an important component of household waste and represent a valuable contribution to soil fertility. Manure contributed by cattle is also important. The symbiotic relationship between farmers (who contribute crop residues as forage for cattle) and herders (whose animals contribute to soil fertility) is an intrinsic part of the social as well as the ecological balance.

On lands that lie farther from the compound, such intensive husbandry decreases and fallowing of gona increases somewhat. These factors combine to result in a strong pattern of crop choice in relation to the location of gona fields. Fields closest to the residence receive the most attention, being devoted primarily to cereal food crops; at more intermediate distances, grain legumes--primarily groundnuts for sale--become mixed with cereals; in the most distant fields, cash crops, especially nonfood crops such as cotton, become relatively more dominant (Norman 1972).

Mixed cropping. Mixed cropping[12] is commonly practiced on gona
land. Gona crops are grown on ridges, generally one meter or pace
apart, by using systematic planting patterns (Figure 3.2), which
permit specific spatial arrangements of as many as 6 or 8 crops on
the same field. In the Zaria area, for example, only 23 percent of
the cultivated upland was sole-cropped in 1966-67 (Table 3.4). As
many as 178 mixtures of crops were identified on the remaining 77
percent of the area, although by far the commonest mixture involved
only millet and sorghum. That combination and 10 other mixtures, in
fact, accounted for 64 percent of the total cultivated rainfed area.

FIGURE 3.2
Spatial Arrangements of Two Common Crop Mixtures, Zaria Area, 1966

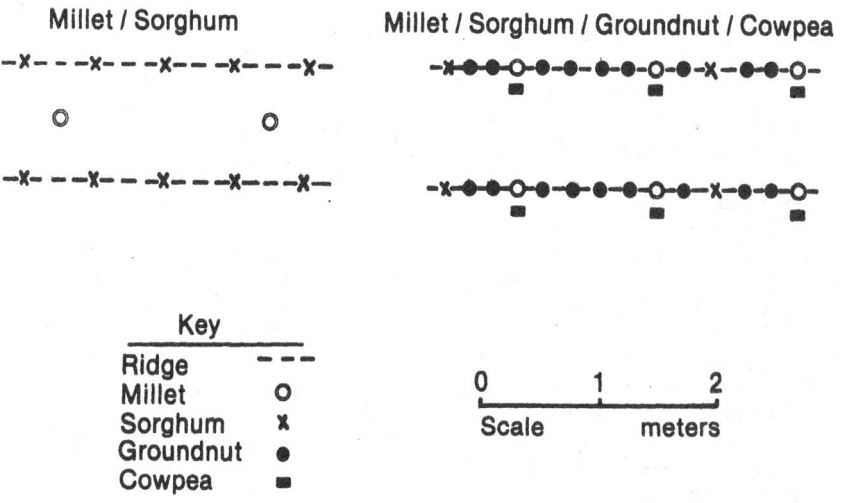

Millet / Sorghum   Millet / Sorghum / Groundnut / Cowpea

Key
Ridge   – – –
Millet   O
Sorghum   X
Groundnut ●
Cowpea   ■

0   1   2
Scale   meters

On the other hand, fadama crops, which are usually grown on the
flat and not in ridges, were cultivated most often as sole
crops--partly because, as the major fadama crop, sugarcane does not
easily permit such mixed cropping. Sugarcane grows in tall, dense
stands, which shade surrounding areas so heavily that other crops'
growth is seriously limited.
    Common gona mixtures, however, generally do not compete for sun
or space to such an extent, and, indeed, it appears that certain
combinations are chosen because from a technical viewpoint they have
complementary biotic relationships. In addition, many crop mixtures
have complementarities from the management viewpoint of the farmers.
    Differing growth cycles of crops are the most apparent
technical reasons for growing crop mixtures; millet and sorghum
mixtures in the Zaria area illustrate this well. Millet is planted

TABLE 3.4
Major Crop Enterprises Grown on Average Zaria Farm, 1966-67[a]

Enterprise	Hectares	Percent of Cultivated Area
Sole crops:		
Sorghum	0.24	7.5
Groundnuts	0.05	1.6
Cotton	0.18	5.7
Sugarcane	0.09	2.8
Other crops (14)[b]	0.16	5.0
	0.72	22.6
Two crop mixtures:		
Millet/sorghum	0.75	23.5
Sorghum/groundnuts	0.08	2.5
Cotton/cowpeas	0.11	3.5
Other crop combinations (45)	0.31	9.7
	1.25	39.2
Three crop mixtures:		
Millet/sorghum/groundnuts	0.15	4.7
Millet/sorghum/cowpeas	0.11	3.4
Cotton/cowpeas/sweet potatoes	0.13	4.1
Other crop combinations (47)	0.31	9.7
	0.70	21.9
Four crop mixtures:		
Millet/sorghum/groundnuts/cowpeas	0.16	5.0
Other crop combinations (38)	0.20	6.3
	0.36	11.3
Five and six crop mixtures:		
Combinations (19)	0.16	5.0
	0.16	5.0
Total:		
Cultivated	3.19	100.0[c]
Fallow	0.75	
	3.94	

[a]Apart from sugarcane, which is a fadama crop, the only crop enterprises specifically included by name are those used in comparing sole and mixed crops (Table 3.5).
[b]Figures in parentheses denote the number of other crop enterprises in that class.
[c]The total number of hectares enumerated amounted to almost 397 hectares.

with the first rains, stays in the field for about 110 days, and is harvested just when the sorghum, planted a week or two after the millet, begins to grow vigorously. Millet's rooting habit also complements that of sorghum (Andrews 1974). Mixing cowpeas with millet is another way of meshing different growth cycles. In addition, in that unsprayed cowpeas are quite vulnerable to damage by <u>Maruca testulalis</u>, there is some evidence that planting them in combination with other crops reduces insect damage (IAR 1972). Mixed cropping, combined with the ridge-cultivation system, permits the adoption of an implicit (within-field) rotation. At the beginning of some years, the ridges are split (Echard 1964; Buntjer 1971), which means that the crops grown on a particular patch of soil will vary from year to year, although the same mixture may be present in the field as a whole.

With mixed as opposed to sole cropping, fields possibly are better protected against soil erosion. When quick-growing and slower-growing crops are combined, the soil surface is covered with foliage for a longer part of the year. In addition plant-population densities in total tend to be higher for intercropped mixtures than for sole stands (Norman 1974).

In response to our queries, farmers themselves did not articulate technical reasons for practicing mixed cropping. The major reasons they cited had to do with returns, specifically returns involving their most limiting factors, land and labor. Farmers also noted the need for security of yield as a major reason for mixed cropping, and many mentioned that it was traditional to grow crops in mixtures. With survival both a traditional and a contemporary goal, there is a certain amount of congruence in these reasons.

How justified are the reasons given by farmers? The results in Table 3.5[13] provide some site-specific information to illustrate their view that growing crops in mixtures is both more profitable and more secure than growing them in sole stands.

This profit maximization view, especially with regard to the use of factors other than land, is, of course, more congruent with an economist's perspective than with an agronomist's. Labor is the farm household's major variable input into savanna agriculture. The average annual labor input per mixed-crop hectare in 1966-67 was 62 percent higher than the input per sole-cropped hectare. Yet, in the peak season for labor use (June and July), labor was clearly more efficiently used on mixed crops; the differential was reduced to 29 percent in that season.

Sole-stand crop yields per hectare were generally higher than those for the same crop grown in mixtures--where yields showed decreases of from 11 to 30 percent. Possible reasons for the lower yields included competition with other crops in the mixture for water, light, and nutrients, and the lower plant-population density of an individual crop when grown in a mixture.

To clarify the significance of the yield data, however, we combined the yields of individual crops grown in mixtures and expressed them in terms of a common denominator. Using that method, we could readily compare the returns farmers realized from sole- and mixed-cropped hectarages.

The superiority of crop mixtures for improving the returns to the most limiting factors was confirmed by the results. The gross margin (Table 3.5) per hectare was 60 to 68 percent higher for crop mixtures, depending on how labor was costed. In looking at the return to labor, the gross margin per annual man-hour expended on mixed-cropped fields was the same as that from growing crops in sole

TABLE 3.5
Sole and Mixed Crops on <u>Gona</u> Land, Zaria Area, 1966-67[a]

Variable	Sole Crops	Crop Mixtures	Percent Change From Sole to Crop Mixtures
Labor (man-hours/hectare)[b]:			
Annual	362	586	61.9
Labor peak period (June & July)	122	158	29.5
Yield (kg/ha):			
Millet	-	366	-
Sorghum	786	644	-18.1
Groundnuts	587	412	-29.8
Cowpeas	-	132	-
Cotton	213	189	-11.3
Gross margin (N/ha) with labor:			
Not valued	36.79	59.48	61.7
Costing hired labor only	33.41	54.02	61.7
Costing peak labor only	30.57	51.42	68.2
Costing all labor	18.33	29.29	59.8
Gross margin (N) per:			
Annual man-hour	0.10	0.10	0.0
Man-hour during peak period[c]	0.20	0.24	20.0

[a]The figures in this table are not weighted equally by village but are a pooled sample of the observations in the three villages. The weighting system used in calculating the labor inputs and yields involves weighting the different enterprises according to their relative contribution to the total area under sole or crop mixtures (Norman, 1974).
[b]These include field work only. Since the productivity of labor, depending on the task, varies according to age and sex, different types of labor were expressed in terms of a common denominator, man-equivalent. The weighting system involved is explained in Table 5.3.
[c] Labor inputs outside the peak period were costed.

stands. That was because the annual labor input from growing crops in mixtures was higher than for crops in sole stands. When labor applied during the labor-bottleneck period was considered separately, however, the return per man-hour during that period was 20 percent higher for crop mixtures. Mixed cropping, therefore, not only alleviated the labor bottleneck in physical work terms, but also paid off in terms of returns to that limited seasonal labor.[14]

Finally, turning to the security criterion which farmers cited, the results indicated that growing crops in mixtures gave a more dependable return (Norman, Pryor, and Gibbs 1979). That was not surprising because different crop species in a given mixture are likely to respond differently to variations in weather and daylength, and to insect and disease attacks. As a result, failure or partial failure of one crop can sometimes be counteracted by compensatory growth by another. The indigenous cropping systems with their emphasis on mixed cropping therefore appeared to be well attuned to the social and economic environment as well as to the physical environment. A balance was achieved between the goals of profit maximization and security by maximizing yields subject to the physical environmental constraints of a limited growing season. Finally, the mixed cropping systems appeared to be well adapted to the relatively low soil fertility characteristic of the area.

Strategies of Agricultural Researchers

Agricultural scientists, on the other hand, have, in the past, devoted little attention to the traditional practices. Instead, they sought to design modern methods of cultivation and management to increase production potential dramatically. The focus generally has been on manipulating certain technical factors--water, soil, and pests and diseases--with an emphasis on modifying those factors to fit the crop or the animal, rather than on altering crops or animals to better fit the technical environment.

Water. Short of irrigation, researchers have suggested that savanna farmers can make the best use of existing rainfall in upland cultivation by planting in a more timely fashion--closer to the beginning of the rains. A major advantage of power cultivation, either by animal traction or by tractors, is that it makes tillage operations easier and faster. In terms of physical properties, mechanical plowing has a beneficial effect in aiding root growth and penetration by increasing total soil porosity. That, in turn, improves infiltration and permeability and, therefore, increases the amount of available water. If done at the right time, such plowing and tillage operations can effectively reduce surface runoff and erosion; if undertaken in the dry season, however, they can exacerbate wind and water erosion because at that time of year, soil aggregates are easily destroyed and the soil will be turned into hard clods or fine dust. As a result, soil compaction likely will occur in the rainy season and permeability will be adversely affected. Similar problems will arise if the tillage operation is carried out when the soil is too wet. Nevertheless, the main benefit of power cultivation is that the plow can be used for basic tillage operations to modify the soil structure and to generate a positive yield response (Poulain and Tourte 1970; Charreau and Nicou 1971; Charreau 1974a), resulting from cultivation at a greater depth than is possible with the use of hand tools and from a greater efficiency of water use.

Soil fertility. As has been noted above, nearly all crop nutrients apart from carbon can be obtained from the soil, but that does not ensure maximized yields. Loss of available soil nutrients results from soil erosion, runoff, leaching, and crop removal. Obviously, to maintain soil fertility or productivity, such losses must be counter-balanced by such factors as cultivar, climate, and soil-nutrient status as well as by cultural practices that influence the degree of nutrient removal by crops. Although little can be done to reduce the nutrients withdrawn in the economic products of the crop, total losses can be minimized if the nutrients contained in the noneconomic products are returned to the soil. Traditionally, in hand-cultivation systems, residues that could not be incorporated into the soil were burned resulting in a continuous loss of nitrogen and sulfur. To agricultural scientists, the inability of farmers to incorporate residues has thus resulted in recommendations to apply purchased inorganic fertilizers.

Kowal and Kassam (1978), however, suggest that the return of crop residues should receive more attention by scientists in view of the potential advantages of such residues: conservation of nutrients, improvement of soil physical properties, and control of runoff and erosion. They also stress that because of the increasingly critical nature of soil fertility, a more integrated approach is needed to ensure its maintenance. That would involve adopting the concept of basal soil and maintenance fertilization, as developed in the francophone countries of the savanna (Charreau and Fauck 1970; Chaminade 1972; Morel and Quantin 1972). The concept involves applying a basal or initial dose of fertilizer to correct soil-nutrient deficiencies, particularly phosphate, and to bring the soil closer to its potential fertility; then manipulating the basal fertility by maintenance fertilization. Basal fertilization should be adapted to the nature of the soil, not to the nature of the crop; maintenance fertilization, on the other hand, should counter-balance all nutrient losses caused mainly by crop removal and by leaching (when erosion is kept under control). With that approach, fertilizer application essentially is moved from a short-run function (where rates are based on the return obtained from the crop being grown in the year of application) toward a long-run perspective.

Plant characteristics. The miracle of high-yielding varieties of crops is clearly the result of scientific success in adapting plant characteristics to certain soil, water, and climate conditions, although it should be noted that getting those conditions right is crucial. Genetic manipulation has been a part of the agricultural-research agenda in the savanna for decades. Short-stalked sorghums and improved varieties of groundnuts and cotton have resulted from such efforts. But despite substantial maize yields, achieved under very nontraditional practices, the savanna has not had the breakthroughs for rice and maize varieties in evidence elsewhere.

Pest and disease reduction. Entomological work on reducing pest and disease attacks through direct interventions has resulted in a number of recommendations, many uneconomic. Seed dressing for sorghum and cotton and sprays to reduce cotton borers, however, have been tested extensively and show promise for reducing losses in yields due to pest or disease infestation. As economists at the Institute for Agricultural Research, we were involved in testing some of the scientifically based innovations that may have significant impacts on future yields (see Chapter 7).

THE PERSISTENT YIELD GAP: SOME THOUGHTS FOR THE LONG TERM

In this chapter, we have focussed on the agroecological conditions and constraints with which farmers and agricultural scientists working in the savanna must reckon. There seems to be an increasing interest among researchers in learning from the farmers' cultivation practices; and farmers are, albeit in limited ways, beginning to use improvements suggested by the researchers' results. Yet they still have a long way to go before the "green revolution of the savanna" becomes a reality.
Table 3.6 gives, at three levels of technology, examples of typical yields for a few of the major crops grown in the Northern Guinea Savanna zone. Is it possible or indeed desirable for experiment-station yields to be obtained by farmers there under practical farming conditions? Let us briefly examine the reasons for the yield gaps, before considering the question directly.

TABLE 3.6
Actual and Potential Yields of Major Crops in Sole Stands, Northern Guinea Ecological Zone, Northern Nigeria[a]

	Indigenous Practices	Improved Practices at Farmers' Level	Experiment Station
Millet	366[b]	1000	2500
Maize	-	2900	8000
Sorghum	786[b]	1530	3500
Cowpeas	132[b]	1534	1700
Groundnuts	587	1229	2300
Cotton	364	784	1500

[a]The figures were derived from various studies and reflect relative orders of magnitude rather than being comparable in absolute terms. Some of the experiment station yields were based on discussions with technical scientists.
[b]The crop is usually not grown in a sole stand under indigenous conditions and therefore the figures reflect those resulting from growing it in a mixture.

The gap between experiment station and farmers' average yields can be attributed to a combination of two major sets of factors:

1. Technical environmental differences between the experiment stations and farmers' fields--something in the technology which is not transferable to farmers' fields even under ideal circumstances--may account for some of the difference. For example, striga problems may be avoided on experiment-station soils due to their higher inherent soil fertility (unfortunately, often the case, according to Byerlee et al. (1981)), rotational systems that avoid the build-up of striga, and other factors. On the other hand,

striga may be an endemic problem under village-farming
conditions and beyond the capacity of one or a few farmers
alone to solve. Little can be done, at least in the short
run, to close this part of the gap between
experiment-station yields and potential farm-level yields.
IRRI, in their rice constraints studies, termed this gap
Yield Gap I (IRRI 1977).

2. Differences in the quantity, level, and timing of such
inputs as varieties, fertilizer, pesticides, and labor can
account for some of the gap between experiment-station
yields and those under practical farming conditions. The
underlying causes for differences in potential farm-level
yields and those obtained by farmers, called Yield Gap II
(IRRI 1977), may be somewhat more complex than might seem
evident at first glance. The differences can be broadly
classified into two parts: first, those that would have
been overcome if the improved system had been correctly
applied by farmers (e.g., the right seed variety with the
right planting density with correct levels and timing of
fertilizer application); and second, those that are
independent of the recommendations for an improved system
(e.g., lack of water, or soil with a particular deficiency).
The latter problem is difficult to overcome, although closer
specification of the conditions under which the particular
improved systems would be applicable and classification of
farmers and farmers' fields according to these conditions
might help. Such a task would involve commitment by more
research and extension services to tailor more closely the
recommendation for specific environments. The reasons
behind the first problem are also complex and are likely to
be strongly linked to the socio-economic environment in
which the farmer operates (as we discuss in the next
chapter). Simplistically, however, this part of the yield
gap is explained by such factors as: technical inputs such
as seed and fertilizer not being available when required;
not being available at prices farmers can afford; not being
practical for farmers to use them in an optimal manner; not
being compatible with the farmers' goals; or not fitting in
with the farming system the farmer is practicing.

Thus, to return to the basic question raised earlier in this
section--even if it were possible, would it always be desirable for
farmers to achieve the yield levels obtained under experiment
station conditions?

The answer would at first seem to be strongly positive.
Development of improved technology to increase agricultural
production is essential if the present farming population is to
survive in agriculture in the long run and/or if any measure of
national food self-reliance is to be achieved. But formal research
programs have long aimed at improving the productivity of crop and
livestock enterprises, emphasizing the modification of the natural
environment to fit the crop or animal in the short term. Yields of
crops have been increased by adding fertilizer, applying chemical
treatments to guard against insects and diseases, and the like.

Animal productivity often has been based on nutritional studies emphasizing what was desirable for the animal; often they bore little relationship to what the natural environment is likely to provide. Now, however, limited development funds and increased costs of fossil energy, combined with the increased realization that the technologies developed have not been adopted by many farming households, are leading to a reappraisal of the approach--and to an attempt to match "what is possible" with "what is desirable", to define what is desirable not only in terms of what is possible but in terms of what makes most sense in the long term. Government programs, such as those for subsidized fertilizer, for example, may be able to improve the possibility of farmers' adopting fertilizer recommendations in the short run and even closing part of the yield gap. But in taking such action, the government has to decide whether it is in the long-term interests of both farmers and society at large to encourage such adoption. One element in this decision is economic: limited development funds raise difficult questions concerning criteria for their allocation. Another element has to do with survival of the natural environment.

The reduction of the amount of land fallowed and the shortened periods of time that fallowed land is allowed to lie idle already have been noted. Both are largely due to increasing population pressure, and as a result increase the problems of maintaining soil fertility or productivity over the long term. As we mentioned above, nitrogen applied in fertilizer can be leached rapidly during the humid period--an inefficient use of an expensive input. Scientists have, therefore, recommended split applications. Given labor constraints, farmers find that, with hand-tillage methods, managing such split applications is difficult. Scientists have begun to recognize the reality of this constraint, along with the low levels of nitrogen and the losses through leaching, and have begun to do research on incorporating plant residues immediately after the end of the growing season (Fauck,Moureaux and Thomann 1969). Organic residues, which are poor in nitrogen, assimilate the mineral soil nitrogen into microbial tissue and release it slowly as the tissues themselves deteriorate. It is likely that such a mechanism could have a long-range beneficial effect as well as short-run impacts on yield. Kowal and Kassam (1978) suggest that the practice of frequent return of crop residues under continuous cropping may, in the future, be accepted as good farming practice and as an essential part of the improvement not only of soil fertility but also of the soil physical condition.

In summary, increasing attention, we believe deservedly so, must be focussed on modifying the biological constraints to production through such approaches as changing the physiology of crops and integrated pest management to enable the crop or animal to fit the natural, physical, or technical environment--rather than modifying the physical environment to fit the crop or animal. We do not mean to imply that the more traditional approach is completely invalid. What we are suggesting is that, in light of the realities, continuing to drastically modify the physical environment surely has severe limitations. Increasing costs of fossil energy demand that energy should be used more sparingly and with greater attention to its efficient use. A greater knowledge of the production environment in which farmers are operating provides greater complexities in terms of the research process, but it is of paramount societal importance for both researchers and governments to develop strategies--even if they involve substantial financial commitment--to preserve the natural resource base for the use of society in the long run. Although we share the view of Johnson (1972) that farmers themselves are researchers and may eventually devise methods compatible with the new resource ratios, these ratios are changing too rapidly for that to occur to save the environment for posterity. The process we are advocating for addressing the yield gaps and the investment for the future is, of course, the farming systems approach to research.

NOTES

1. In northern Nigeria, Kowal (1970a; 1970b) found that, during the humid period, any continuous rainfall of more than 20 mm contributed significantly to erosion.
2. This advantage is now being diluted because of the increasing numbers of irrigation schemes.
3. This supports the finding that the potential yields of maize are much higher in the savanna areas than in the wetter areas farther south--where maize has been grown traditionally (Kassam et al. 1975).
4. Average daily temperatures in the Zaria area, for example, vary monthly from 22°C in January to 29°C in April.
5. Quinn (1974), for example, showed that the yield of tomatoes was heavily influenced by the date of planting, which in turn reflected the thermal regimes. Tomatoes are particularly sensitive to a relatively narrow range of day and night temperatures. As a result, tomatoes planted during the January-to-May periods, when day and night temperatures tend to fluctuate the most, gave unsatisfactory yields; those planted outside that period provided high yields.
6. The so-called C-4 crops such as sugarcane, maize, sorghum, and millet benefit particularly from the regime that exists at this time. C-3 crops such as cotton, groundnuts, and rice also benefit (Kowal and Kassam 1978). C-3 and C-4 refer to different ways in which carbon is fixed, which is important in the photosynthesis process (Black 1971).
7. The clay that is present is predominantly kaolin, which has a relatively low cation-exchange capacity.
8. The base- or cation-exchange capacity indicates the quantity of exchangeable cations that a soil can absorb. It is expressed in milliequivalents of cations per 100 grams of soil or of clay or colloid.
9. The buffering capacity indicates the potential of a soil to resist appreciable pH changes. This property is directly related to the soil's content of colloidal material--clay and organic matter--and of carbonates, phosphates, and similar compounds.
10. Much more detailed discussion is available elsewhere (Charreau 1974a, 1974b and 1978; Jones and Wild 1975).
11. That is flowering is not dependent on seasonal changes in

daylength.

12. We define mixed cropping as the practice of growing two or more crops on a given piece of land at the same time. The different crops may be together for a short or a long time. Such a characteristic has made an acceptable definition of crop mixtures a contentious issue. For our purposes, any degree of overlapping in terms of time is considered to be mixed cropping. Shortness of the rainy season precludes double cropping--sequential cropping on gona land. However, on fadama land, sequential cropping is practiced to a minor extent.

13. Although not verified by direct measurement there appeared to be no significant differences in the soil fertility of land devoted to sole and mixed crops.

14. Linear programming models using the same data verified the superiority of mixed cropping under indigenous technological conditions (Ogunfowora and Norman 1973).

4
Farming Communities and Institutional Arrangements in Northern Nigeria

"Successful development projects must take account of community organization and power dynamics. Without strong support, interventions that disrupt the status quo are unlikely to succeed."

Lewis (1955).

The Hausa people who live in the northern Nigerian savanna commonly live together in clusters--towns, villages, and hamlets. Many family groupings reside within calling distance of each other and a fair amount of daily interchange takes place. In the days of wars and slave-raiding, such clusters provided protection and security. Although villages rarely have visible walls today, the remnants of old mud fortifications can still be seen in some places.

In the dry season rural villages seem to be extensions of the bare earth in the empty fields surrounding them. In the rainy season villages just a short distance off the roads are hidden from passers-by by the lush greenery of millet and sorghum (guineacorn). Two to three meter walls built of mud, guineacorn stalks, or grass mats define the residences; compounds or gidaye[1] provide shelter, protection, and security within. The zaure or entrance room built into the wall is usually of mud, as are the one- or two-room structures (daki) inside the walls where people sleep and store their personal belongings and the round storehouses (rumbuna) where the compound's supplies of grain are kept.

Galvanized iron has replaced a few of the mud or grass roofs in many villages; for some time, cement has been a bit more widely used than formerly to fortify mud walls and to pave parts of the compound interior. New public buildings--a school, a mill where the grain grinding engine is located, and market stalls--not only are iron-roofed, but also may have windows, painted walls, and cement floors. Still, in the dry season, the brilliant red-orange of peppers spread to dry on domed thatch roofs vividly contrasts with the monochrome of the lateritic brown of walls and roofs. Stacks of green-brown bundles of groundnut haulms balanced on roof edges add a new height to the low village skyline.

The Maguzawa, Hausa but not Muslim, live in more scattered homesteads and compounds, normally with only one family grouping per dwelling. Settled Fulani households have joined the Muslim Hausa households in villages, whereas nomadic Fulani continue to migrate between villages and the "bush". In the rainy season, the possible competition between Fulani cattle and growing crops, as well as the availability of forage in the open grasslands of the savanna bush, keeps the distance between nomads and settlers fairly wide. In the dry season, however, nomadic encampments of portable grass and leaf huts spring up close to the outskirts of mud-compound villages.

The seasonal transition in the spatial relationship between nomadic herders and settled farmers has acquired a regular rhythm over the decades. Settled people need milk to drink and manure to increase the fertility of their fields. Nomads need grain and vegetables for themselves as well as fodder for their livestock. The flow of mutual benefits from this generally amicable relationship is, however, easily threatened. Population growth and urbanization in northern Nigeria, as in other developing countries, have over time gradually reduced the areas of open land available for wet-season grazing. Growth and expansion of urban areas have caused the nomadic herds of the settled Fulani to move farther and farther from the village in search of forage. The ritual of the seasonal transition itself may be in transition.

The Zaria villages and the many others like them are undergoing changes that are visible, but also other types of changes are underway. The traditional agricultural production unit, the gandu, is breaking up; the acquisition of fertilizer requires that cooperatives be formed; the little girls are beginning to go to primary school along with the boys and are learning the ABCs. The route to Mecca is no longer a two-year trip overland; village women now fly to Jiddah and return within a month as Alhajiya--and with bracelets, holy water, radios, scarves, and slippers for their friends.

In this chapter, we look at how national and state endeavors to spur development are affecting some of the community and family structures. Sometimes the community or household changes; sometimes the development initiatives die because they are founded on ideas unknown or unacceptable to the communities for which they are intended.

THE PEOPLE: VILLAGES, COMPOUNDS, AND HOUSEHOLDS

Though no two villages or households are alike, some generalization is necessary to begin to understand the relationship of the savanna environment to the villages and the households within them and to understand the socio-economic organizations to which people belong. Rural people live in groupings related to the practice of farming as well as to the ties of marriage and kin. There are two broad types of groupings: gandu and iyali[2] (Buntjer 1970a).

Gandu organization implies that there are two or more adult men, one or more of them married, jointly operating a common set of fields. The production process is generally supervised by one of them; as in a father-son gandu, for example, where the father is the chief decision-maker. In some cases, a more mutual decision-making process may exist; as in a brother-brother gandu, for instance. Gandu organization also implies that the dependent kin of the active adult male members of the gandu (wives, children, old parents) all eat together, at least for the evening meal. That means that the wives of the gandu members share cooking responsibilities, with each woman generally taking a turn in rotation, and in cooking use the produce from the common fields. The gandu head is responsible for selling any excess produce and also for purchasing common needs--additional food items for the shared meals, for example. The members of gandu may also cultivate individual fields (gayauna), over which they have individual control of both inputs and outputs as well as their own application of labor.

Iyali organization implies that the grouping includes only one adult man and his dependents. In some cases, an iyali group closely resembles a nuclear family, but polygamous families also qualify. Nephews, nieces, grandchildren, and grandparents are often members of iyali groupings; if nephews or sons are big enough to work on the farm but not old enough to marry, the farming organization may resemble a gandu situation but still technically be thought of as an iyali.

Compounds are physical rather than social or socio-economic entities, as are gandaye and iyalai. Entrance to the compound is generally gained only through the zaure or entrance room. Non-kin men are rarely permitted beyond the zaure, as women are inside the compound and are not to be seen by strangers. The head of the compound (mai gida) generally controls access; the zaure serves as his public room, and during the late afternoon and evening hours of leisure, several male visitors usually occupy the mats and sheepskins placed for seating on the zaure floor.

Women in major parts of northern Nigeria, including the Zaria area, keep various degrees of purdah (kulle), or seclusion, in the compound. In the more remote villages, and, apparently, in the traditional urban centers, the practice is kept more strictly and women confine their infrequent visiting of friends to evening hours. In our study village closest to Zaria (see Chapter 5), however, the settled-Fulani-female population rarely stayed inside, although stranger-men were still not permitted to enter the compound where women wore clothing less modest than that worn when they went out. The Fulani women explained their behavior in terms of their responsibilities as milk-sellers, a traditional nomadic Fulani women's task, rather than in terms of Islam. Although Hausa women in this village did not have this reason for their behavior, they, too, appeared to step out of their houses during the day somewhat more often than did women in the more remote villages--to remove chaff from grain on a windy hill just outside the village, to attend a nearby clinic with a sick child, or to perform other tasks. Modernization and the example of others in setting behavior standards no doubt has had an influence on this practice.

Women in villages occasionally work at farm tasks outside the compound, more often as hired laborers in the harvest season than otherwise. Generally, our study revealed that women who did such work were poor, or did it as a favor for male relatives, or had an occupation (sana'a) for which the crop being harvested was a major input (such as cotton for a weaver).

Each compound may contain one or more iyali or gandu. Sometimes the internal space of the compound is physically divided by walls or other barriers; in others, there are simply two or more cooking fires in the open space, each used by a different consuming unit. As defined in most Nigerian surveys, including those we were associated with (see Chapters 5 and 6), household or family relates to those who cook together as members of a "pot"--"suna ci daga tukunya daya." The consuming unit, or pot, in the villages seemed to be closely related to the farming unit (gandu or iyali), but only partially related to the compound, or gida.[3]

The village head (sarkin gari), as the appointed authority in charge of land allocation and the assessment and collection of taxes, plays a powerful role in village affairs. He maintains law and order in the village and may adjudicate or arbitrate local civil cases. Officially recognized as the village's representative in external affairs, and paid an annual salary by the Local Authority to discharge these duties, the village head's approval is essential for outside agencies to establish contacts with individuals or households within the village. The village head's household, therefore, is generally included in research endeavors as well as in sanitary inspection and other official contacts involving villagers and strangers.

The village head has the authority to appoint other individuals to village offices, including those of hamlet heads. Often assumed in such an appointment is a personal or clientage relationship that involves obligations and rewards between the appointer and the appointee. Thus, the village head, because of that authority within the village as well as his broker role with the outside, has, if so inclined, considerable opportunity to enrich himself. His obligation to maintain harmony in the village, however, acts to curb wanton exercise of such opportunities. Still, it is unusual to find a village head who is not a member of the economic as well as the socio-political elite.

The village head is appointed by the district head, who is appointed by the Emir or Sultan. This hierarchy is both Islamic and traditional. The colonial rulers in northern Nigeria chose to use the administrative structure established after the Jihad in 1804 by Usman Dan Fodio, Sultan of Sokoto. Thus, modern and traditional concepts of civil authority are to some extent mixed.

The state government, headed by an elected governor, is part of the federal system of governance in Nigeria. The state government delegates a certain amount of authority to the appointed Emir/district head/village head hierarchy while maintaining other authority in the permanent, professional civil service hierarchy (ministries, district offices, etc.) and the system of modern--as opposed to Islamic or shariya--courts. Thus, the appointed Emir is responsible for administering a number of districts; appointed district heads (often public civil servants) have councils composed of traditional title holders responsible for communicating government mandates to the villages and supervising their execution. Major administrative tasks are the collection of taxes, construction of public works, recording of statistical information, and the execution of government ordinances on subjects as varied as sanitation and primary education.

Whereas the village head may assist in carrying out some of these responsibilities, district heads also call on the civil service structure for technical assistance--the Ministry of Public

Works, the Ministry of Agriculture and Natural Resources, and the like. Where his area is involved, the village head is expected to communicate with the ministry personnel, perhaps to organize a cooperative or to line up voluntary contributions of labor or cash from the villagers. In short, the village head is expected to play two roles: that of a modern administrator and that of a traditional leader.

COMMUNITY NORMS AND BELIEFS

The norms and beliefs shared by the people in savanna villages in northern Nigeria would, if they could be enumerated, run into the tens of thousands. Here we focus on a few of those that affect the potential for change, particularly for agricultural change. Individuals' decision-making processes are rooted in their understanding of and adherence to societal norms and beliefs. Where individuals perceive that a given action may bring personal gain but threaten security or status within the community, they may refrain from acting. Community norms thus at times exert powerful deterrent forces against innovations and individual initiatives. Indeed, in the development literature, traditional norms are often pictured as villains encouraging conformity and blocking progress. On the other hand, community norms and beliefs foster a certain amount of harmony and accord by establishing a common understanding of acceptable behavior and thus guidelines for a daily life. And, as we discuss later, beliefs do change over time; norms do get modified and revised.

The almost complete acceptance of Islamic tenets constitute perhaps the greatest source of social and cultural influence in much of northern Nigeria. The institution of purdah, the naming ceremonies for eight-day-old children, the organization of the day around the times of prayer, the practices of tithes, charity, and alms (zakka and sadaka), attitudes toward schooling and intellectual life, and the relationships among men, women, and their children can all be related to the practice of Islam. But there are equally pervasive cultural and social patterns not necessarily related to religion: the organization of the households, sense of pride in adulthood, the choice of crops and diet, a myriad of friendship and kinship interaction patterns, the ambition of women to establish a degree of economic independence, the assignment of traditional titles and roles, the customs of marriage and child-rearing, a sense of fatalism.

We look here briefly at those norms that appear to affect the work that people do and how they do it, the roles that children play, and the social interactions between households involving time, goods, and money.

Cultural/Societal Norms and Work

Rural men are, by and large, farmers in their own right. Though land is in theory owned by the village as a whole, in fact it is also commonly bought, sold, rented, loaned, and inherited by individuals. Under the communal traditions that prevail in much of Africa, land is thought of not only as a factor of production, but

also as a significant element in the social fabric of the community
(Dunsmore et al. 1976). Polanyi (1964) observed that "land is a
tangible dimension of the community and is that part of nature that
is interwoven with man's political institutions." The link of land
with the community entrusts it to "a vast family of which many are
dead, a few are living, and countless members are as yet unborn"
(Elias 1962; Uchendu 1967). In essence, land provides community
stability, continuity, and a basic prerequisite for rural work, and
the role of land becomes more significant rather than less in a
period of rapid social and economic change. Economic, religious,
social, political, and historical variables collectively provide the
definition of and the context for change in the status of land in
African societies.

Land tenure in many parts of the West African savanna has a
double ancestry: in the traditional concepts of communal ownership
and in the tenets of Islamic land law. The communal land laws give
people usufructuary rights to the use of land within their own
communities (Abalu and Ogungbile 1976). Legislators in northern
Nigeria have passed laws consistent with this concept, granting
ownership of all land to the government, and, in turn, sanctioning
the continued allocation and control of land at the community level.
At that level, the representative of government is the village head
(Goddard 1972; Oluwusanmi 1966). Under Islamic land law, by
contrast, individual tenure is recognized (Goddard 1972) and rules
have been established governing personal inheritance. The passing
of rights from generation to generation at the village level tends
to follow the Islamic code, although village heads do have a right
to intervene. In general, however, with patrilineal inheritance
systems, the use rights of land have been handed down from fathers
to sons, with a sense of private ownership and control being
developed over time. In many Islamic lands, daughters also have the
right to inherit from their fathers; because women in northern
Nigeria generally do not farm, they are encouraged to surrender
those rights and in practice only men make claims to farming land.
Women's frequent marriage outside of their own communities also
makes this a practical course of action.

In theory, no individual has the right to alienate land (or the
use rights of land) from the community to which he belongs, but that
has in fact happened upon occasion (Hill 1972). Again, in theory,
the government would place no bars to alienation of land from users
for reallocation to others; in practice, when land acquisition has
been necessary--as in the establishment of major dams and irrigation
systems--the government has tried to provide use rights in
equivalent land or in the improved land to the former users.

The availability of additional cultivable land is sometimes a
difficult question. Costs of clearing bush in some areas are so
high that the creation of additional fields appears to be an option
open to those already land-rich rather than to land-poor farmers
(Hill 1972). In other areas, free bush for expansion of fields
simply does not exist. Still, only those rural men who have too
little land to produce a household subsistence and too little
capital or too few skills to do any other occupation will accept
full-time work as laborers for hire on others' farms. That may be
changing, however, as higher wages are offered for labor on

irrigation-development and similiar schemes or as the pressures on land become excessive and other low-capital occupational options are closed down. Rural men have responsibilities to provide at least two meals a day for their dependents and to provide new clothes for them once a year--usually at the festival of Sallah, Eid el Fitr. If farming cannot provide this minimum, other work will be sought.

Rural men traditionally have learned their occupations, especially farming, through on-the-job training on their fathers', uncles', or other male relatives' farms. Thus, a rural man who learned the cropping patterns of a successful uncle knows that the seeds for a certain crop are to be planted, for example, a pace apart. What similar socialization process does an extension agent offer? Instead, a farmer is told that, with fertilizer, one pace is too far. Similarly, a rural man knows that a successful man should have as many children and wives--up to a maximum of four wives--as he can afford and to keep his wives in purdah. What changes in perspective must he acquire to restrict the number of children--who provide his future labor force--or to reduce the number of wives to which he aspires?

Only rural men who are Fulani, whether settled or nomadic, know about cattle. But only settled Fulani men treat cattle as an investment rather than an occupation, for they appear to be able to find trustworthy nomadic Fulani who will herd them on loan (riko). Hausa men keep goats, sheep, and donkeys, but rarely cows. This stereotype regarding knowledge about and work with animals will have to be overcome if farmers--both Hausa and settled Fulani--are to use ox-drawn plows and cultivating equipment.

Rural women, by and large, are wives, mothers, and small-scale entrepreneurs. Married by the age of thirteen or fourteen, and normally bearing their first child one year later, girls are socialized to become women very quickly. After marriage, they are also, most of the time, found inside their compounds working not only at the expected household tasks but also at a wide variety of independent economic activities. Virtually all women do such work (sana'a) to earn money. However, this simple economic explanation incorporates a complex mix of social as well as economic motivations toward the accumulation of an independent store of wealth. Three aspects of the male-female relationships and the division of roles within the household help to explain why women seek to acquire independent financial resources.

First, while the male head of household is largely responsible for care and maintenance of the household, women are expected (by their husbands) to provide for their own personal needs such items as soap, cosmetics, room decorations, and some clothing. Women must also supply dowries for their daughters, particularly enamel and brass pots, clothing, and room decorations. Nearly 90 percent of the 212 rural women interviewed in one survey further indicated that they provided at least part of their own midday meals and those of their small children, as well as such items as snacks, kola nuts, and cigarettes (Simmons 1976c). And in three cases, women actually supplied the means of sustenance for their entire households.[4]

Second, men spend their daylight hours working or visiting with men outside their compound, while women remain inside with other women and children. Women, from childhood and through later kin and marriage ties, develop independent social contacts with other women, frequently with women outside their compounds, whom they visit on special occasions or after dark. These friend (kawa) relationships may take different forms, but a gift relationship, which is more or less prescribed, generally appears to be an essential part of these friendships and elicits certain financial responses from one's friends. Thus, the moral support offered by a confidante outside the compound helps to motivate a woman to earn money, to cement and maintain the friendship.

Third, women also try to be good providers for their children, for when they are old and widowed, a strong parent-child relationship may serve as women's only form of social security. Sons who can provide housing for old mothers are especially valued. Friction between parents, however, often results in the mother's leaving the compound without her children, who remain behind to live with their father. If a mother has trained and treated her children well, providing gifts of food, clothing, and money, they may feel obliged to provide for her if she later requests assistance.

Women's motivations to work are also influenced by social norms and conditions other than monetary needs. Women believe that a married woman must have an occupation (sana'a) to establish herself as a respectable adult in the community. Newlywed young girls, old women who become weak or senile, and women new to the village are virtually the only women allowed to be idle. There are many acceptable excuses for interrupting one's working life, but it is said that a woman should be shunned by other women if she is able but unwilling to engage in some independent economic activity. Although they recognize that the erosion of demand for crafts, particularly for hand-spun cotton thread, has added an element of risk to the choice of a money-earning activity, women still voice the opinion that all women have some opportunity for remunerative work.

The feeling also seems to be prevalent that a women should not depend too strongly on her husband, reflecting perhaps the ease with which men can divorce their wives and the fact that one wife in a polygamous marriage is more expendable on practical grounds than a single wife in a monogamous union. Polygamous marriages offer distinct advantages to women; they also present organizational problems that can work to one woman's disadvantage and render her position less secure.[5] Working and saving a portion of one's earnings--often by buying small livestock--are hedges against insecurity, in that independent financial resources reduce dependency both on one's husband and on the smoothness of one's relationships with co-wives.

Cultural/Societal Norms and Children

Rural children, if they survive the first two years of life, are socialized into their adult roles and the rigors of savanna agriculture fairly rapidly--girls faster than boys. They generally begin by tending younger siblings and half-siblings, helping to collect firewood, selling ready-to-eat food items for their mothers, and carrying water. By the age of about six, children are expected to be able to fend for themselves--taking care of their own clothes, finding additional food if they are hungry, and making regular work contributions to the household tasks.

Male children are allowed a great deal of freedom in their nonwork time; female children seem to have more extended duties at home and are somewhat less free than boys to go about with friends. By twelve years of age, girls are being prepared for an arranged marriage, while boys do not begin to think of wedding until they are capable of assuming a full workload in farming and earning cash.

Koranic schooling traditionally has been more important for boys than girls, although girls do attend school on occasion. Prior to the recent introduction of the scheme for Universal Primary Education, most villages had no primary schools, and only boys were allowed to go to schools outside the villages. Girls were generally kept at home to work. When a primary school is available in a village, however, girls as well as boys attend--when their mothers can spare them and their fathers (or the mai gida) approve.

Secular schools have often been considered as a sure means of escape from the village, especially for boys. The reluctance of parents to permit their children to attend the schools thus may be based on their wish not to lose the labor of their children in the long as well as the short term. Further, the potential for children to receive cultural and social values inconsistent with community norms and beliefs cannot be overlooked. Some parents' reluctance to let their children attend secular schools is possibly more strongly related to this factor than the work factor. Alhaji Junaidu (1972), a respected Islamic theologian, has eloquently discussed the societal dangers resulting from the development of such conflicts.

Cultural/Societal Norms and Community Interactions

Households' social and cultural interactions are more than the sum of individual relationships among members of households, such as women's kawa relationships, already noted, and the routine visiting among men. Family and kinship ties also bind whole households in a network of mutual obligations and rituals. One observer noted that the exchange of bowls of cooked food among households caused nightly traffic jams on the town's footpaths even though the food exchanged was often very similar! Marriages, funerals, and naming ceremonies are also major causes for expressions of mutual support and account for a substantial part of the leisure time of individuals.

Three Islamic customs--zakka, sadaka, and the feeding of almajirai (Koranic students) --provide the rationale for significant inter-household and inter-personal assistance of goods and cash. Zakka is a tithe; farming households at harvest time separate a certain amount of the output for redistribution to those who are less fortunate (Hays 1975a). The village head apparently acts as a middleman in some cases. In others, the gift is direct. Sadaka is a gift; beggars, old people, and gift-bearers are given sadaka as a matter of religious obligation. Normally, sadaka is directly transmitted in small amounts from one person to another. After mosque on Friday, men often regularly dispense a few kobo for sadaka to waiting supplicants. Women often give sadaka in kind, taking a piece of ready-to-eat food from that which they are selling and "dashing" it to the person in need. Food for almajirai is a special kind of sadaka; it also is a matter of religious obligation--for both givers and receivers who by begging learn humility and dependence. In rural villages, however, where the Iman (liman) has a farm, his resident scholars may work on it and, in effect, become household members. In other cases, the scholars may each bring substantial foodstuffs, usually grain, from home and only supplement their staple through begging. In urban areas, the sight of begging almajirai is common and many residents regularly feed a number of these students.

Failure to live up to social and religious obligations may be a matter of shame, but doing exceedingly well at them does not seem to be a matter of inordinate pride. Being overly generous indeed may bring disapprobation, as it comes close to the negative value of flaunting one's wealth. Earning status through extraordinary generosity is rare.

SOCIETY IN TRANSITION

The large-scale political and economic changes that have altered the national image of Nigeria since independence in 1960 have also affected the lives of people in villages and hamlets across the savanna. The changes that are taking place at present are so profound that it is difficult to envision what village life in Nigeria will be like in twenty years' time. Visible manifestations of change are readily apparent. Among the most striking are the construction of roads, improvement of communications, hence movement of products and people to and from villages, and the building of primary schools which, as implied earlier, are having short-term and possibly long-term repercussions on the lives of families in villages.

Among those changes that may be even more significant, but which are not so readily perceived, are those taking place in the relationships: between the individual and the family, between the individual and the community, and between the family and the community.

Changing Family Structure

The breakup of the gandu system of household organization is the most apparent reflection of changing approaches to the task of farm production. Whereas historically the gandu was the more prevalent, and preferred, mode of family organization, the iyali is rapidly becoming the norm. In the survey of three villages in the Zaria area, for example, only 49 percent of the households employed the gandu system in the early 1970s, (Norman, Pryor, and Gibbs 1979). Similar evidence from other parts of northern Nigeria and indeed throughout the West African savanna confirm this trend.

Several reasons have been suggested for the breakup of gandaye: the introduction of cash crops (Reboul 1972; Nicolas 1960); secular education and the influence of the Islamic land-inheritance rule (Venema 1978); increased off-farm employment opportunities (Sutter 1977); the presence of certain views on personal and family relationships (Buntjer 1970a); and the opportunities offered by new settlements and migration. However, the speed at which this transition from complex household organizations such as those of gandu to a more simple mode of organization (iyali) takes place may be tempered by the strength of the traditional hierarchical structure, ethnic origin of the people concerned (Pelissier 1966), ownership of cattle (Buntjer 1970a), and specific farming systems and their labor requirements (Netting 1965).

Perhaps the most important impact of this transition with regard to agricultural development is that the demise of the gandu will encourage further off-farm and cash-earning behavior on the part of farmers whose farms are insufficiently large to provide for family needs. Insofar as farms are large enough to generate substantial cash surpluses, their management by the head of an iyali or a gandu should not make much difference except, perhaps, with regard to the employment of hired labor. A young farmer was perhaps more secure starting out in a gandu because he had a longer period of apprenticeship in which to learn both farming and decision-making skills. Responsibility must be assumed more quickly by a young farmer who begins a farming career and a family at the same time.

The Individual and the Community

Just as young farmers are increasingly reluctant to accept the decisions of their fathers or gandu heads unquestioningly, and to surrender personal wishes to family needs, so too there has been an increasing tendency to question the traditions of community authority. The formal system of administration and rule at the village level has long implied that decisions are to be made by the few for the many. The wishes of the majority, therefore, might or might not be heard. Communication between the governed and their rulers is most efficient from the top down. The route for feedback on decisions from the bottom up is often closed or limited to personal communication between friends or between clients and patrons. As Damachi and Siebel (1973) have suggested, in that social status allocation in Hausa society traditionally has not been based on achievement, the leadership would have to be receptive to change to retain authority. Although village leaders generally have prescribed limits to their actions, they might, for explicit reasons, resist accommodation of personal interest if community interests were threatened.

Times are changing, however. Until relatively recently, most community organizational patterns were designed to ensure collective survival. Now, even authorities find themselves tempted by personal profit. With the construction of a tomato paste factory in Zaria, for example, the technology for growing tomatoes with irrigation during the dry season proved to be particularly profitable. As a result, in some villages existing ownership or use-right agreements on the very limited amounts of lowland (fadama) suddenly were abrogated by those in authority in order to reallocate the land and, thereby, the benefits of the technology to themselves (Agbonifo and Cohen 1976). As in that example, to the extent that the authoritarian pattern of control is used to divert resources from those who have no influence with those in power to those who do, inequities and conflict can result.

Village opinion still plays a very important role in

prescribing and proscribing behavior, however. The decision of individuals, however influential, is not enough; when the pressure of village opinion turns against them, they may be forced to reject ideas that led to their personal gain in order to retain community approval.

The directions that a particular change may take are often found in a range of already acceptable community behaviors. There is generally an accepted community norm, as we suggested in the earlier parts of this chapter, but there are also accepted variations in this norm. For example, it is usually said that in the Nigerian savanna, women do not work on the farm but instead stay in compounds while men make farm decisions. With the major exception of cattle-owning Fulani households, that statement would be widely accepted as a true description of a community norm. Yet harvest of cotton and peppers often depends on women's labor; many women help with groundnuts, too. A decision to extend the area of these crops without planning for this critical seasonal labor might well run into trouble. Further, if we extend the concept of farm work to include the processing of farm commodities, then most women could have substantially greater roles in farm decisions than field-labor figures will show. Crop varieties that produce in abundance but cannot be readily processed by hand methods are likely to be accepted by women only if they are provided with cash so they can use the mechanical grinding-mill services.

Though it is often useful to paint a picture of community behavior with the widest possible brush, it should be recognized that doing so poses the risk of overstating conformity to norms and understating the potential for change.

Changing Roles of Family and Community Labor

Traditionally, with gandu structures and large supplies of unused land in the village environs, junior members of households had little incentive to offer their labor for hire outside the gandu. Where labor was short, either seasonally or for community activities, a method of providing reciprocal labor on an unpaid, communal basis (gayya) served well. In recent decades, however, there has been a demise of organized group labor and a growing trend toward individually hired labor (Raynaut 1973; Unité d'Evaluation 1976). The trend has been evidenced in reports of significantly greater employment of the more productive modes of non-family labor involving cash payments for work, either kwadago (at daily wage rates) or jinga (payment when a task is completed) (Norman, Pryor, and Gibbs 1979).

In addition to a possible weakening in the sense of community obligation, declining household sizes associated with the breakup of gandu have contributed to this trend. Family labor forces are sometimes no longer sufficient to get the farm through the bottleneck-weeding period from June to August. And with farm sizes declining as populations grow, some farmers with very small farms find themselves in need of cash to supplement their subsistence incomes. They thus offer their labor to meet other farmers' demands, even though that may mean neglect of their own farms at critical periods, with consequent loss of yields (Matlon 1977).

Thus, both the nature of the way that nonfamily agricultural labor is recruited and used and the availability of such labor are changing (Kohler 1968; Monnier et al. 1974; Raynaut 1976; Ernst 1976). It would seem that cash progressively is becoming a more important factor in a farming-system operation as well as in the consumption patterns of households. Whether the market orientation in the labor market preceded or was the result of market orientation in production cannot be said with any certainty; that such an orientation is likely to have profound implications for further agricultural development and for socio-cultural change in community life is a certain conclusion.

EXTERNAL INSTITUTIONS

Private trading systems are perhaps the most well-known and longest-lived institutions that link villages to the outside world. Traditionally, the marketing of produce over long distances was in the hands of private individuals and companies. After the colonial government instituted state intervention in trade in the early part of the century, the Federal Government of Nigeria continued the practice, albeit for different reasons. The government marketing boards for cotton and groundnuts--both a source of innovation and a source of exploitation in the past--continue to function in an atmosphere of contention today, although over the years they have been restructured and reorganized many times.[6]

In recent years the federal government has also assumed a more interventionist role with regard to direct agricultural investments. More financing of rural infrastructure and credit services, expansion of personnel participating in the development projects (often funded by outside donors), and establishment of a range of national agricultural institutions characterize this new role of the federal government. The most important new institutions as well as the older ones are listed, along with the purpose of each institution and references for evaluating their achievement of goals, in Table 4.1.

Agricultural planning, the eighth institution listed in the table, is currently the task of the Federal Ministry of Economic Development (FMED). The FMED's role, and that of its Central Planning Office, has been mostly that of loosely monitoring state plans (prepared by the state Ministries of Agriculture and Natural Resources or Planning Ministries) and guiding them in the direction of federal policy. Limited assistance is given to the states on information and a data base, project development, and budgeting. The state planning ministries are responsible for carrying out these tasks, but limited manpower has prevented their doing them adequately. Given the states' limited capacities and the federal interests in demonstrating agricultural initiatives, planning emphasis has thus been on irrigation projects, often large-scale ones such as Kadawa in Kano State. In recent years, increasing attention has been directed to improving rainfed agriculture, largely through the mechanisms of the World Bank supported Agricultural Development Projects (ADPs). All of these factors have led to an underemphasis on creating a structure for broad-based agricultural growth.

TABLE 4.1
Some Major Specific Agricultural Related Institutions in Nigeria[a]

Institution	Type[b]	Origin	Function	References
1. Extension Service:	S		Main delivery system for dissemination of information and advice to farmers	Buntjer (1970b) Buntjer (1972)
Operation Feed the Nation (OFN)	S/I	1976	Broad approach to increasing food production involving the whole population	
National Accelerated Food Production Program (NAFPP)	S/I	1970's	Coordinated technological package approach to provide inputs to increase production	Edache (1978) IITA (1977)
Agro-Service Centre (ASC)	S/I	1976	Provide agricultural inputs and advice	
Agricultural Development Project (ADP)	S/I/0	1974	Integrated agriculture development projects which ensure availability of all farmers' needs	Huizinga (1978) D'Silva et al. (1980) D'Silva and Raza (1980)
2. Agricultural Inputs: Fertilizer Units	I		Provision of fertilizer for distribution to states for distribution to farmers	Falusi (1973) Laurent (1969) Falusi and Williams (1981)
National Seed (NSS)	I	1970's	Production and distribution of seeds	FAO (1970)
3. Agricultural Credit: Nigeria Agricultural Bank (NAB)	I	1926 1970's	Provision of seasonal and short term loans	King (1976a)
4. Agricultural Cooperatives: Produce Marketing	S/I/0		Marketing especially of the export crops	King (1976b) King (1975) King (1978)
Thrift and Credit	I		Provision of savings mechanism and seasonal loans	

5. Marketing Board:	0	1947	Marketing of first export crops and more recently food crops. Boards now exist for cocoa, cotton, groundnuts, grains, palms, rubber, tubers, and roots	Olayide et al. (1974), Abbott (1974), Adamu (1970), Akintomade (1974), Helleiner (1974), Kriesel (1974), Titiloye et al. (1974)
6. Agriculture Development Corporations:	S/I/O	1960's	Wholly owned state government agencies responsible for developing large-scale plantation agriculture and other commercial ventures	
7. Research Institutes:	S	1920's	Organized along commodity lines on national basis	Idachaba (1980)
8. Agriculture Planning: Central Planning Office Federal Ministry of Economic Development (FMED) State Planning Ministries	S	1960's	Coordinates federal and state planning in its Central Planning Office. Generate information and a data base for project development and budgeting	Mijindadi (1976), Raay (1975), Simmons (1971)

[a]This is not a complete list but simply reflects those receiving major emphasis at the present time.
[b]Type of institution code: S = service; I = input; O = output.

In this section, we briefly look at the range of external institutions, both private and public, that serve to link the villages of northern Nigeria with a network of markets for inputs (fertilizer, seeds, and advice) and for products (grains, export crops, and legumes). We recognize the vital importance of many institutions that exist outside the village and that broadly affect rural development progress inside the village--such as transportation, communication, education, and other public services--and the policies that guide them, but we focus here only on those institutions geared primarily and most directly to the incentive structure for agriculture.

The Input Markets

Agricultural inputs generally include fertilizers, chemicals, pesticides, herbicides, fungicides, and seeds. Credit and agricultural advice may also be considered as special kinds of inputs, however, and we do so here.

Fertilizer. Many approaches to the fertilizer-distribution process have been tried; all have met with little success. The result has been shortages and inadequate coverage (Agricultural Planning Division 1974; Falusi 1973). It has been estimated that the rate of fertilizer consumption amounted to about 3.6 kg/ha cultivated for the main crops in 1969, compared with the average minimum requirement of 89 kg/ha cultivated (Agricultural Planning Division 1974). The main problem is providing fertilizers, rather than a lack of need or demand for them. Experimental and demonstration-trial results show that it is profitable to apply fertilizer on many crops. In a recent study in Nigeria on fertilizer distribution, it was found that institutional factors are a more important determinant of the level of fertilizer use than is price (Falusi 1973). In fact, supply unavailability and lack of adequate working capital were found to be major restrictions on the expansion of fertilizer sales. The major shortcomings of the present system have been:

1. The failure of states to acquire an adequate supply of fertilizer at the correct time (i.e., early before planting season) because of bureaucratic inefficiency, failure to initiate purchase tenders in time, default on contracts by those receiving the purchase order, and other reasons.
2. Because it may arrive late and there may be lack of transport or other organizational problems, the fertilizer often is not dispensed to local store centers for distribution on time.
3. The consequent shortage of fertilizer at planting season often has resulted in private agents charging more than the subsidized rate for fertilizer, which has resulted in the states wanting to stop using private agents.
4. There has been little coordination of fertilizer policies among states; consequently, prices have varied, resulting in scarce supplies moving across state lines to take advantage of higher prices.

5. The amount of fertilizer that ultimately gets distributed is small partly because of the above factors and partly because of the lack of credit to allow farmers to purchase fertilizer.

Seeds. Organized quality-seed production and distribution in the past were limited to cotton. Only recently have attempts been made to expand the production of other types of improved seeds. A National Seed Committee now coordinates the activity of the recently created National Seed Service (NSS), which is responsible for producing and distributing foundation seed to the state Ministries of Agriculture and Natural Resources (MANRs). These operations are just beginning, but to date the multiplication and distribution of improved seeds has been inadequate. It has been estimated that in the late 1960s less than one percent of the farmers were planting improved varieties of food crops (FAO 1970).

Credit. Agricultural credit formally originated in 1926. Since that time, despite numerous attempts to make credit available to the small-scale farmer, most have not been successful. The Nigerian Agricultural Bank (NAB), established in 1973, has had only slightly greater impact than local commercial banks in expanding credit to small farmers. Agricultural cooperatives in various forms have long existed in Nigeria, but they have achieved only limited success in a few localities.

Institutional credit has been linked to programs aimed at introducing mixed farming--that is, integrating livestock and crops. Initially in the Mixed Farming Schemes, oxen were to plow the land and to provide a source of manure (Alkali 1969). In more recent attempts to overcome the labor bottleneck at land-preparation time, government-subsidized Tractor Hire Units (THU's) have been used; credit has been an integral part of these operations, which have had mixed success (Weber 1971; Purvis 1968).

Extension advice. The federal government served predominantly in a research and extension role until the Second National Development Plan (1970-74), when it moved into a more direct role of investment and policy intervention in the agricultural sector. As we discussed earlier, research results and extension advice have been coordinated through extension-research liaison services. The extension service is intended to disseminate to farmers information and advice related to improving agriculture through talks, demonstration plots, agricultural shows, and other techniques. The service can be characterized, however, by its low level of extension concentration (i.e., one extension agent to every 2,000-3,000 farmers) and, until recently, by its attempt to serve the whole of Nigeria uniformly. That has had an extremely diluting effect; in none of the nine study villages in which the Rural Economy Research Unit worked in the Hausa area of northern Nigeria were any extension agents active or even known.

Other extension-related programs, such as Farm Institutes to train prospective farmers, also have had only limited success (Olukosi 1976). Therefore, the continuing problems of the extension service and its inability to serve the whole farming community in recent years have led to a number of different programs, with services tied to the delivery of specific technical packages.

Foremost among these programs are the National Accelerated Food Production Program (NAFPP), Operation Feed the Nation (OFN), Agricultural Development Projects (ADPs), and the Agro-Service Centers (ASCs). The National Accelerated Food Production Program is a cooperative venture involving all food-crop research institutes, the state Ministries of Agriculture and Natural Resources, and the Federal Ministry of Agriculture and Rural Development. NAFPP's objective is to help increase food crop production by introducing high-yielding varieties, fertilizers, pesticides, and other key inputs through a coordinated technological package. Operation Feed the Nation is even wider in scope and is involved in programs to get the whole population to have backyard gardens to increase food production. The National Accelerated Food Production Program led to the creation of the Agro-Service Centers, which are to provide both inputs (such as improved seed and fertilizer) and advice to farmers.

The central idea of the Agricultural Development Projects is to transfer already developed crop and mechanical technologies in package form to the majority of the farmers in the project areas. The technologies are demonstrated to the farmers by extension workers based at the Agro-Service Centers. In addition, distribution of inputs, especially fertilizer, also takes place through the Agro-Service Centers. The projects began in 1974-75 in three different areas, with the intent to expand the number of areas.

The Output Markets

The vast majority of Nigerian agricultural produce moves through networks of private markets. Direct government intervention in the product movements has been largely confined to operating the marketing boards for exportable commodities. The operation of these boards was altered in 1977, when the entire marketing system was reorganized. Separate boards were created for marketing cocoa, cotton, groundnuts, grains, palm produce, rubber, and root and tuber crops. Tobacco is a notable exception; its marketing is controlled by several large companies. Much literature is available on past performance and problems of the marketing board system in Nigeria (Onitiri and Olatunbosun 1974; Helleiner 1974), so rather than dwell on the details of that system here, we turn to the workings of the private market system.

Types of private markets. The most simple form of private-sector trade is that which takes place between households at the village level or at some local exchange point such as a roadside station. The next most complex form of marketing occurs where people meet periodically in some organized manner to buy and sell goods to satisfy their needs as well as to exchange information with relatives, friends, and strangers. Beyond these rural markets are the larger daily markets found in urban areas. Virtually every farmer participates in some way in this market hierarchy.

Simple exchanges of goods within the village occur daily. Many people trade in goods from their houses; others sell to and purchase from mobile traders (talla), who walk through the village paths and compounds hawking their products; still others regularly set up small tables along the roadsides to sell small amounts of produce--often fruit in season, kolanuts the year-round--to passersby.

FIGURE 4.1
Marketing System Showing Possible Distribution
Channels for Food Crops, Zaria Area, 1970-72

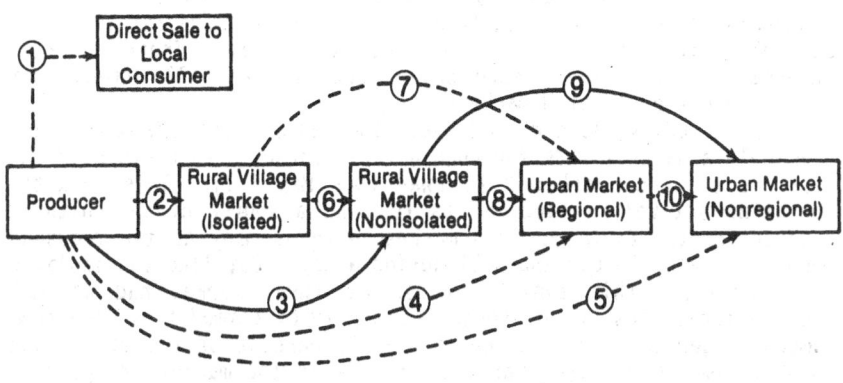

———— Denotes major rural-urban link
– – – – – Denotes other possible links

Periodic markets are also well-used by village residents. Each village or cluster of villages has a market once or twice a week. These markets can be classified as to their isolation or accessibility to motor transport (Figure 4.1). Rural markets, which are isolated (that is, inaccessible to motor traffic), serve most village and local community needs. The most common modes of transport to these markets are by foot, bicycle, and donkey. But lack of accessibility does not mean that these markets are not well attended. One rural isolated market, for example, held once a week, was attended by 1,200-1,500 people on an average market day. More than 75 percent of those attending came from within an 8 km radius and 8 percent came from more than 16 km away. Market attendees both bring products for sale and usually make some purchases. The transactions are often accomplished through village retailers, who receive a small commission for bringing buyers and sellers together. In addition to these participants, even itinerant traders selling nonagricultural goods are found in isolated rural markets. They usually arrive on bicycle and display their wares only on market day. Storage facilities at such isolated markets are minimal, so little produce is stored. Most of the commodities sold in the market come from nearby areas and excess supplies are returned home with the seller. Occasionally, a local trader will purchase or assemble produce in isolated rural markets for transport to, and sale at, another isolated rural market or a more accessible rural market. But, in general, very few products are moved from these markets into the chain that ends in the urban communities.

Accessible rural markets also serve the villages and local communities where they are held, but because they are located on or near a motorable road, they serve in addition as focal points for collecting products that ultimately will be transported and sold in major urban areas. In fact most agricultural/food commodities destined for urban consumers pass through the accessible rural markets, which usually are held only once, sometimes twice, a week. They draw both farmers and local traders, many from distances of 16 to 24 km. A typical day at an accessible rural market will find producers, local assemblers, transporters, and village retailers all actively participating in product exchange. Their greater accessibility also means more buyers and sellers will attend them than attend the isolated markets.

In a study village called Doka, located on the Zaria-Kano road, seemingly a natural site for such a rural market, residents used the services of Sundu market 2 km away (see Map 5.1). This market's operation actively reflected its role as a source of agricultural produce for the Zaria regional market, with as many as ten thousand people arriving to buy and sell during a day. But like the isolated rural markets, this typical accessible rural market had minimal storage facilities, even though the supply of commodities came from a much larger area surrounding it. Sixty percent of the millet and sorghum came from beyond an 8 km radius of the market; 25 percent came from more than 32 km away. Once products reach such an accessible rural market, they are sold through a village retailer in much the same way as in an isolated rural market. Rural assemblers hire truck transporters to convey their purchases to some larger urban market or consuming center. In this particular accessible rural market, the loading and hiring process was so extensive that one person was occupied full-time with coordinating and supervising the operations.

Regional urban markets are those in relatively large cities located in the immediate geographical regions producing the crop, whereas nonregional urban markets serve areas of the country where a particular product is not produced. Obviously, a market would be classified as regional if one concerned with marketing were moving a particular locally produced crop and as nonregional if monitoring the marketing of a crop imported into the area from another producing region. Zaria's main market, Sabon Gari, for example, is a regional urban market for sorghum and millet but a nonregional urban market for the palm oil, oranges, rice, and gari coming from southern Nigeria. The distinction is important because moving products from regional to nonregional urban markets, as from Zaria to Ibadan, usually involves at least one more marketing intermediary than moving within a region.

Urban markets are the major source of products and other goods for urban consumers. Though supermarkets and shops in Zaria do provide alternative supplies of certain products, relatively few consumers use them regularly--particularly if the products are locally produced. Once products reach urban markets from one or a series of rural markets, normally arriving in substantial bulk, they are handled by many wholesalers and retailers. Because the urban market is held often and potentially handles a greater volume than does a rural market, traders who frequent that market acquire more permanent physical facilities than found at a rural market. Most urban market stalls are walled as well as roofed and have doors with locks to provide for secure storage of produce. In that the supply of agricultural products comes from a much larger area than at rural markets, and from many locations, such storage of larger volumes is essential. In Zaria's Sabon Gari market, when we observed it, 24 percent of the supply of millet and sorghum came from within a 40 km radius and 30 percent from more than 160 km away.

The organizational structure just described applies to the marketing of a wide range of agricultural products, although some crops do not fit into any such general framework and so are handled in special ways (Hays 1976). Some cash crops, for example cotton, groundnuts, and tobacco, are marketed largely through specially licensed buying agents and/or the marketing boards. Another major cash crop, sugarcane, which is important in areas of large fadama, is often sold standing in the field, with the purchaser being responsible for harvesting and transportation. In general, however, marketing structure (Figure 4.2) can be distinguished first by the quantities of each product moving through the markets and secondly by the duration of storage at different stages in the marketing channels. Basic staple foods such as millet and sorghum, or any other foodgrain that can be stored by the producer and marketed throughout the year, are likely to move through a sequence of markets. Luxury products, however, such as cowpeas and rice, are more likely to move through types of exchange points other than markets and commonly are stored by intermediaries within the marketing channel. A roadside station or an assembler's house will sometimes serve as the bulking point. Perishable crops are also likely to move through a series of exchange points other than rural and urban markets, especially in that most rural markets are held only periodically. Continually refering to special cases would be difficult, hence most of the discussion and analysis in this section concern marketing two major staples, millet and sorghum.[7]

Transferring the produce. The marketing process can be divided into three stages: first, transfer of produce from farm to rural market; second, transfer of produce from rural to urban market; and third, transfer of produce from urban market to consumer.

As many as three intermediaries in the rural-urban marketing chain can be involved in a transfer of produce from the farm to a rural market. The producer may take his produce directly to the rural market, performing the needed market functions himself. Or he may hire a local transporter to market it.[8] Once in the rural market, the produce can be sold either to retail purchasers or to rural assemblers who will ship it onward. If the produce is to be retailed, it will be placed by itself in a basket or container by a village retailer, separate from other products, to be measured only as the retailer sells it and collects the money. The producer or local assembler who brought the produce to the retailer will receive that money and will pay the retailer a commission in cash or in kind for his selling services.

Transfer from the rural market to the urban market can involve

84

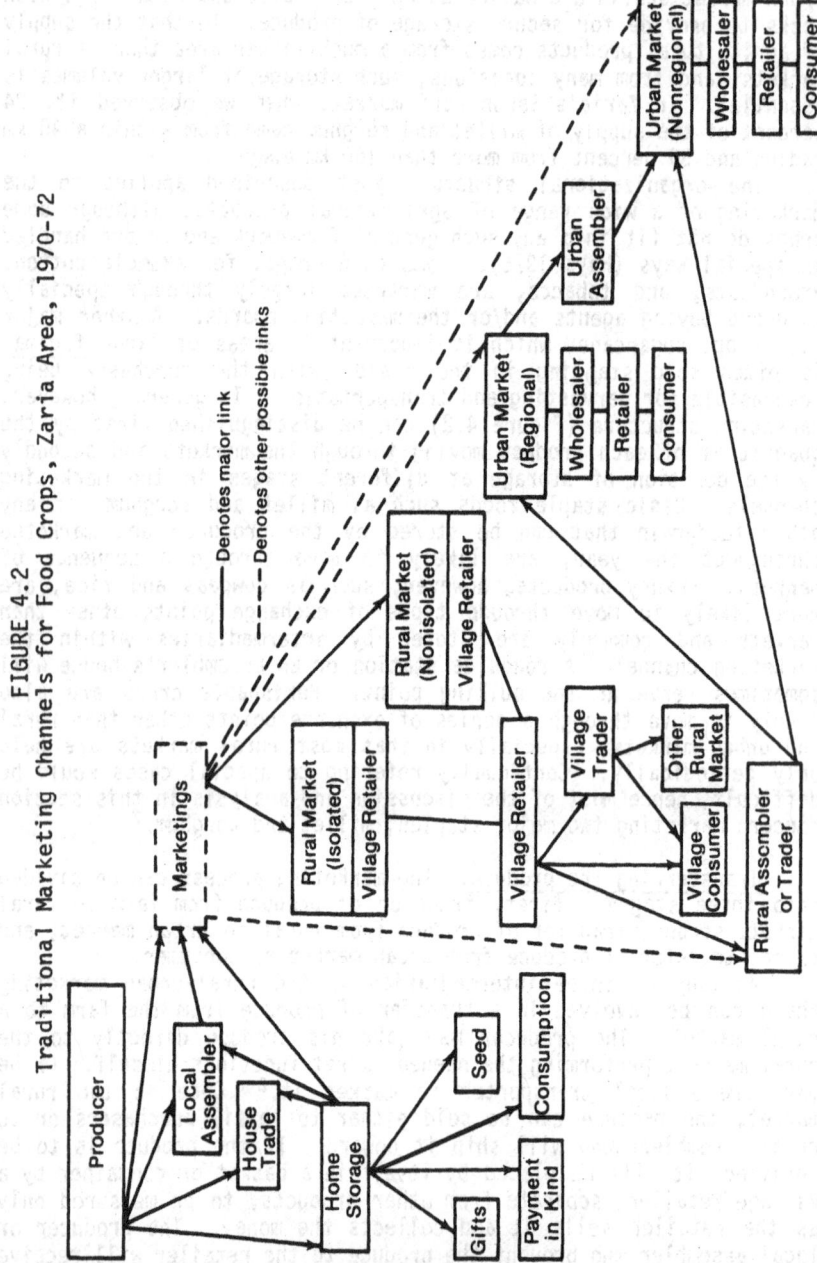

FIGURE 4.2
Traditional Marketing Channels for Food Crops, Zaria Area, 1970-72

—— Denotes major link
---- Denotes other possible links

as many as four possible intermediaries. Village retailers measure and sell produce to rural assemblers as well as to consumers. Rural assemblers supply empty sacks or containers, then hire truck transporters to convey the produce to urban markets. Rural assemblers also pay for produce-loading at the rural market and unloading at the urban market. Transporters generally are paid only for transport services, normally on a unit or volume basis. A market official takes charge of organizing truck loading for transporters, and he receives a fee from the transporter for doing so. Rural assemblers usually ride with their produce to urban markets, where they deal with a wholesaler or commission wholesaler. Soon after reaching urban markets, produce usually is stored in a wholesaler's stall before retail-selling. If the wholesaler is a commission wholesaler, he receives a commission for selling the product from the rural assembler. If he is a regular wholesaler, he retains all the profit on the transaction. Retailers purchasing stocks from wholesalers normally have to pay for transporting the produce from the wholesalers' stores to the retailers' stalls. Retailers display the produce in their own stalls, sometimes after grading the product, and measure it as they sell it. Consumers provide their own containers for their purchases. Figure 4.3 illustrates the major marketing channels used for marketing sorghum and millet in three villages near Zaria.

Marketing functions and margins. The organizational structure of the traditional agricultural-marketing system just described provides the framework within which the pricing system gives expression to the preferences of consumers and guidance in the allocation of resources. The structure directly affects the degree of market competition and the efficiency of price formation.

The marketing intermediaries add to the value of the grain by performing services that require labor, time, and capital and therefore add to farmers' and consumers' costs. The difference between the price that consumers pay and that the farmer gets is the marketing margin. Marketing margins for one rural-urban link in the Zaria area, summarized in Table 4.2, show the share of the average yearly retail price of one sack of millet and of sorghum received by each marketing intermediary. During the year the producer received an average of about 68 percent of the final retail price of millet and almost 70 percent of that for sorghum. There was a close correlation between the producer price and the retail price for both millet and sorghum (Hays 1975a).

Intermediaries perform a wide array of services, some of which require great flexibility in their operations. The absence of any public facilitating programs, such as a market price-information service, means private entrepreneurs must rely entirely on their own initiative and personal contacts for carrying out their operations. Experience and understanding of the local environment underlie their ability to adjust market decisions to provide useful and convenient services. Experience of the intermediaries in the grain trade we studied ranged from an average of six years for local assemblers and transporters to eighteen and twelve years for wholesalers and retailers, respectively.

The price shares received by those groups to some extent

86

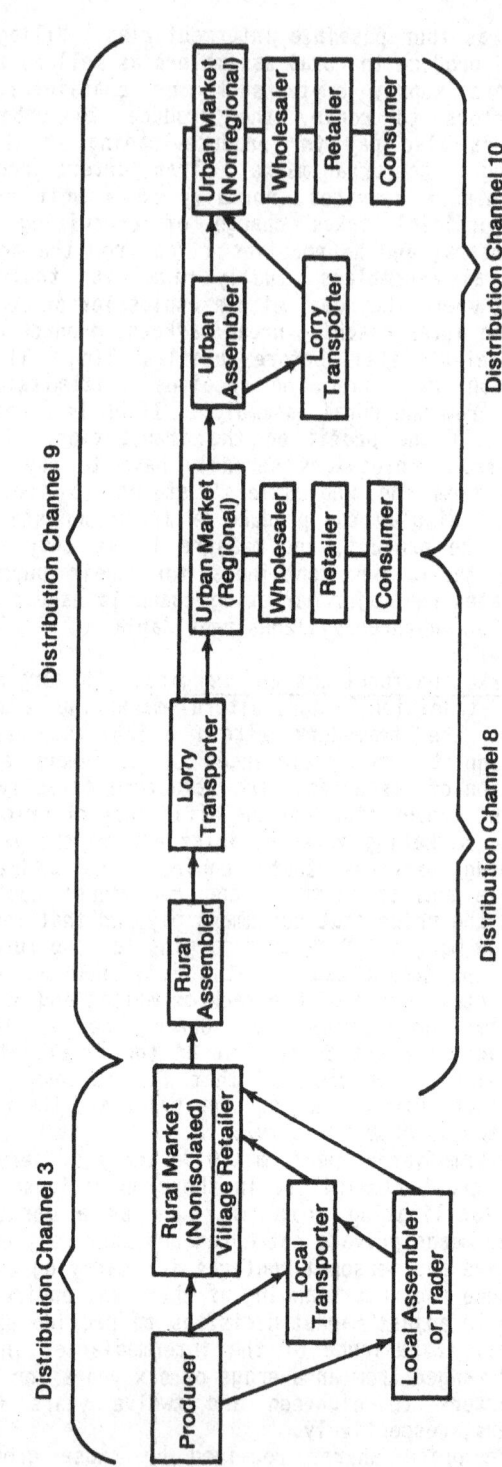

FIGURE 4.3
Producer to Consumer Distributive Process for Food Crops Showing the
Major Rural-Urban Link, Zaria Area, 1970-72

TABLE 4.2
Share of Retail Price for Sorghum and Millet Received by the Farmer
and Marketing Intermediaries, Zaria Area, 1971-72

Individual	Percent Share	
	Sorghum	Millet
Producer	68.2	69.8
Intermediary:		
Local assembler (trader)	9.1	9.5
Local transporter	2.6	2.2
Village retailer	2.0	2.2
Rural assembler	5.0	4.1
Lorry transporter	3.9	3.2
Urban market wholesaler	2.6	2.2
Urban market retailer	6.6	6.8

reflected the differentials, although local assemblers commanded the highest rate for their services. The marketing margins of the various intermediaries remained fairly constant throughout the year.

A functional classification provides a framework in which to examine the nature of each of the intermediary's activities. These functions are grouped into exchange, physical, and facilitating functions (Table 4.3).

Although all seven intermediaries performed important services, some functions were duplicated four or five times and all eleven functions mentioned in Table 4.3 were performed at least twice. At least four of the seven intermediaries each performed eight or nine of the eleven functions. It is likely that the size of operations is critical to the number of functions performed. Most intermediaries have a small volume of trade and low investment and visit only a small number of markets each week (Hays 1975a). That tends to result in a lack of specialization in trade, with under-utilized capacity and minimal ability to introduce innovations and absorb technological improvements. The marketing system does, however, mobilize resources in the form of both skilled entrepreneurship (for example, assembling, buying, selling) and capital (for example, financing, storage, transportation), which would not be available to the public sector for use in an alternative arrangement.

To determine the importance and role of private entrepreneurs in the rural-urban link, we must analyze their businesses and marketing operations. Table 4.3 shows annual returns achieved by the various intermediaries in the Zaria area performing their marketing functions in 1971-72. In addition most of the intermediaries engaged in other occupations, including farming, but we excluded expenditures and earnings from those enterprises. A comparision of the returns of intermediaries together with market-structure information provide insights into positions of market power of the different intermediaries.

TABLE 4.3
Annual Incomes and Marketing Functions Performed by Intermediaries, Zaria Area, 1971-72

Variable Specification	Local		Village Retailer	Rural Assembler	Lorry Transporter	Urban Market		Total Times Functions Performed
	Assembler	Transporter				Wholesaler	Retailer	
Exchange:								
Buying	X			X		X^a	X	4
Selling	X			X		X	X	4
Negotiate			Y			X		2
Physical^b:								
Assembly	Y^c			Y				2
Transport	X	Y			Y		X	4
Storage	X			X		X	X	4
Facilitating: Standardization	X		X				X	3
Financing	X			X		X^a	X	4
Risk bearing:								
Physical	X			X		X^a	X	4
Market	X			X		X^a	X	4
Market information	X		Y	X		X	X	5
Total functions performed	10	1	3	8	1	8	9	
For intermediary:								
Annual return (N)^d	155	33	109	303(405)	3224	271(542)	286(381)	
Number of other occupations	1.4	1.4	1	0.6	0	0	0	

^{a}Not applicable where wholesaler operates as commission wholesaler and does not take title from rural assembler.
^{b}Processing is not included because processing of millet/sorghum is normally done by consumer.
^{c}Y denotes functions unlikely to be provided more efficiently by alternative arrangements.
^{d}Figures in parentheses include income from dealing with other crops besides millet and sorghum.

Average monthly returns ranged from nearly ₦3 for the local transporter to almost ₦248 for the truck transporter. Even after allowing for return to capital, the return to the truck transporter--the only intermediary with any significant amount of durable capital invested--was more than ₦174. Clearly the truck transporter--who does not take title to the grain--is in a different class from the other intermediaries. The truck transporter's earnings in the Zaria area in 1971-72 included all hauls--grain, people, and other goods. Such a high monthly return to labor and management would seem to be in excess of the opportunity cost of transporters' services, although we cannot state that categorically. Were the return in excess of the opportunity cost, the high return would imply that access to capital allows excessive profits. The average charge for transporting grain locally, however, was ₦0.13 per tonne kilometer; on long hauls, charges were about half that. These rates compare favorably with transport charges in other developing countries.

The rural assembler had the next highest income, which would be consistent with the large total number of functions he performs, most of which require considerable entrepreneurial skill. The local assembler had the largest margin of any intermediary. Although his average margin for the year, ₦0.80, for assembling a sack of grain might appear excessive at first glance, it seems less so when the services and their costs are considered. Consider, for example, the services of the local assembler who provides an empty sack, goes about purchasing grain, assembles these small purchases into sack quantities, then either transports or pays for transporting the grain 15 to 25 km to a rural market, waits for it to be sold and then returns home. He provides a sack for the grain; which he assembles, and spends most of a day taking this grain to market; at least a day's labor is involved per sack. The opportunity cost of his labor can be approximated from the wages of alternative employment. Results from the farm management study indicated that average imputed farm incomes were ₦0.84 per man-day family labor on the farm,[9] which would compare favorably with the local assembler's margin.

The village retailer had a monthly return three times that of a local transporter, but considerably below that of other intermediaries. Although village retailers perform important functions in mediating the bargaining process, their services do not require much investment nor involve much risk. Monthly returns for wholesale commission agents and retailers in the urban market were considerably higher, excluding their income from selling other crops. Including income from sale of other crops, commission wholesalers made about ₦1.50 a day and urban retailers about ₦0.80.

Therefore, it does appear that in general the incomes of the intermediaries were not excessive, considering services provided, and there was little evidence that intermediaries were able to exploit inefficiencies in the traditional market structure to increase their share of the final retail price. Generally, intramarket competition among a large number of intermediaries both at the rural-market and at the urban-market level tended to suppress margins and limit profits that could be earned in the grain trade. That leads us to conclude that the rural-urban part of the marketing system appears to function reasonably efficiently within the environment in which it operates. Thus, the marketing margins shown in Table 4.2 reflect both the multiple marketing services and the length of the marketing chain more than the cornering of exorbitant profits on the part of individual intermediaries.

Intermarket and seasonal price relationships. Imperfections in the marketing system could also result from intermediaries taking undue advantage of differences in prices between markets and seasons. To ascertain whether in fact that was so with respect to the basic food staples--millet and sorghum--prices for 1969-71 were analyzed for fifteen selected markets in four northern states of Nigeria.

The results revealed that the price spread was often in excess of transfer costs, implying imperfections in the market (Hays and McCoy 1978). Although there was a high degree of competition within the local subsystem, our analysis indicated a possible lack of competition among subsystems. However, there was evidence that the excessive price differences among urban markets did not result from planned manipulation under monopolistic or monopsonistic conditions. Rather, they were a result of imperfections inherent in the system which, due to certain characteristics of production and marketing, made effective response to intermarket price differentials difficult.

Consider first the nature of millet and sorghum production. There is a lack of specialization in the production and therefore a lack of concentration in supply, with only small surpluses available at many different markets for intermarket trade. A large portion of grain is stored at farms, so the marketing patterns and storage practices of producers are important in determining the supply available at any one time and location. Although some farmers store grain to take advantage of seasonal price rises, probably more store grain because they need cash through the year; and that is an important determinant of timing of disposal. In fact it causes unpredictability in farm marketing, compounded by defects in the marketing system: a lack of adequate information on crop prospects, surplus areas, and prices, and a lack of specialization by traders taking part in trade between markets. All information must be obtained and disseminated through private contacts, as there is no public information available. Along with the nature of production and farmers' marketing patterns, this introduces uncertainty of supply and increases the risk of trade in more distant markets, where there is even less information. This prevents specialization and many small-scale traders develop contacts in certain areas, to keep informed on market conditions, and engage in trade in those areas, with little knowledge of market conditions elsewhere. Markets around centers are competitive, but the network of markets is not integrated.

Using the same fifteen markets, we examined seasonal-price relationships for the same period[10] by calculating the net seasonal rise--the rise above that considered consistent with storage costs--in millet and sorghum prices (Hays and McCoy 1978). The results indicated considerable variation in seasonal price increases both among markets and between months within a given year. Millet is usually harvested in August-September, and seasonal-price movements showed that high and low points were consistent with the harvest period. In all but five instances, however, for the fifteen markets over the two years studied, the yearly average seasonal increase in millet prices exceeded the calculated expected increase. Sorghum is usually harvested in November-December, which corresponded with its low price, but the high prices generally occurred several months before harvest and a little after millet harvest. This was not only because some millet is substituted for sorghum in peoples' diet, but also because by the time of the millet harvest farmers could estimate forthcoming sorghum crop prospects and, if they were good, market their stored surplus sorghum. In all but four instances over the two years studied, the yearly average seasonal increase in sorghum prices exceeded the calculated expected increase.

The net seasonal price increases can be used to make hypothetical estimates of traders' unit profits, depending upon assumptions about the timing of storage decisions. The net seasonal-price increase can be interpreted as a gross return, as its computation allows for all storage costs except for the risk factor and a return to the trader for his entrepreneurial ability. Where the net seasonal rise in price exceeds the expected price rise, there is opportunity for traders to make a higher than normal profit. The extent of those profits depends on the traders' skill in purchasing and decisions on the length of storage. A policy of purchasing at harvest and storing until the off-season high price occurred would not necessarily result in the highest unit profits. The great degree of variability required in marketing to achieve the highest unit profits illustrates that there is a considerable element of risk in storage operations.

The important question involved is whether seasonal-price increases result because traders have monopolistic power to influence prices through their storage operations and thus to earn abnormal profits. The argument in our study indicated only possible profits and did not show whether they were actually attained by traders. Other findings in the study strongly suggested that traders did not have the monopolistic power to attain such profits. Evidence supporting that view included the findings that little storage took place by traders in the urban market; that in urban markets, traders' monthly purchases were about equal to monthly sales; that there was a continuous flow of grain to urban markets from the rural areas; and that a large amount of grain was stored by farmers. To the extent that the rural-urban marketing link in northern Nigeria reflected price changes back to the producer, it was the producers who benefited from the seasonal price rises, an observation that assumed homogeneity among the farming population. In our next chapter we will discuss whether some producers--for example the wealthier ones--benefit from the seasonal price fluctuations while the poorer ones are either forced to sell grain at low prices immediately after harvesting or must, because of the lack of food self-sufficiency, purchase grain later when prices have risen.

NOTES

1. The singular is gida.
2. Plurals of these words in Hausa are gandaye and iyalai; anglicization into gandus and iyalis is heard when people are speaking English.
3. Therefore, whereas the mai gida and his household were consulted about survey participation, their pot was included in the survey itself only if it were selected as a sample household. The village head, however, was always included in the survey (see Chapter 5).
4. This was a matter of shame in one woman's household; her husband would go off to another city to seek work, leaving her with nothing. The women in other households, however, happened to be wealthy in their own right. In one of these, the woman actually employed her own husband as her farm manager.
5. It is no accident that the Hausa words for co-wives (kishiya) and jealously (kishi) are so close.
6. For a discussion of the marketing board system within Nigeria, see Onitiri and Olatunbosun (1974).
7. For an empirical discussion of some other products see Gilbert (1969) and Ejiga (1977).
8. Many relationships between farmers and transporter-traders have fairly long histories. Many factors can change the existing relationships, however. A road constructed nearby can suddenly make a donkey-trader's services less attractive to a farmer who, by carrying his produce to the road himself, can perhaps make a good connection with an urban-oriented commission agent or a cost-conscious truck driver and increase his own share of the value. A need for cash can cause a farmer to shift his business from a low-margin but poorly-capitalized trader to one who takes a bigger cut, but offers short- or medium-term credit. There is of course a danger in the development of such hierarchical trading relationships, particularly with organizations and individuals outside the village. The potential for exploitive relationships developing where trading and credit are linked are obvious (Clough 1977; Watts 1978; Palmer-Jones 1978), although such relationships were not apparent in the Zaria area villages studied.
9. The actual figure in 1966-67 was ₦0.52 (Norman, Pryor, and Gibbs 1979), which in 1971-72 terms--when the marketing study was undertaken--was equivalent to ₦0.83 after allowing for an average annual inflation rate of 10 percent.
10. For an analysis of the seasonal grain price variations for a number of years before the Nigerian Civil War, see Gilbert (1966) and Jones (1968).

5
Farming Systems
in Three Zaria Villages,
Northern Nigeria

"In the (countryside) innumerable bush paths lead from
compound to compound, village to village. These paths
have been maintained through the years by the tread of
feet."

Pedler (1955)

The Institute for Agricultural Research (IAR) and the Faculty
of Agriculture of Ahmadu Bello University (ABU) are housed in a
growing complex of offices, laboratories, and classrooms in Samaru,
about 16 km from the ancient city of Zaria. Once a rural
agricultural experiment station, IAR is gradually being engulfed by
the booming university town which surrounds ABU's Main Campus nearly
3 km away.

The old city of Zaria, founded in the sixteenth or seventeenth
century by Queen Zaria, still serves as the local administrative
center of the Zaria Emirate. The Emir resides in and governs from a
large and colorful compound in the center of the city. But a new
city of Zaria has grown up outside the walls of the old, with two
nuclei on the road between old Zaria and Samaru. It is the
combination of new and old urban functions that gives Zaria its
character and regional prominence. Tudun Wada, one of the newer
nuclei, is a minor commercial center. But Sabon Gari -- literally
translated, new town--is the major commercial, service, and
manufacturing center of modern Zaria.

OVERVIEW OF RESEARCH: THE SETTING

Village Selection

As the process of identifying sites for research got underway
in 1965, the possibility of selecting sample communities in the
Zaria zone of influence was considered. For logistic reasons alone,
such a choice made sense. Further, as the criteria of site
selection were developed, it was evident that Zaria qualified as a
regional market center. Finally, and somewhat surprisingly,
traditional farming practices in the area were far from
over-studied.

The criteria used in selecting the villages in the Zaria area

93

were the same as those later adopted in choosing the study villages in the Sokoto and Bauchi areas.[1] The three criteria were the following:

1. The villages should differ in their ease of access to the regional market center.
2. The intermediate-access village should have a relatively higher proportion of land capable of supporting crops in the dry season.
3. The village heads should be cooperative and ready to support a long-term research relationship.

The first criterion was derived from the concentric ring theory of von Thunen, later reformulated by Schultz (1951). The later version, in which both factor and product markets were considered, was based on the reasoning that farmers' incomes would tend to be higher nearer urban areas because of the greater efficiency of the factor and product markets. Applying that criterion in the Nigerian savanna also led to the presence of another gradient: villages with better access also had denser populations.

The second criterion was specified on agroecological grounds. Its purpose was to capture the differences in farming systems that would evolve when it was possible to extend agricultural activities into the long, dry season. In applying that criterion, the physical environments of potential villages had to be assessed before making a selection. Because sufficiently detailed soils and hydrological maps did not exist to permit that to be done in the office, field inspections were necessary, and they facilitated the application of the third criterion, which involved talking at some length with village heads and soliciting their cooperation.

In all areas, including Sokoto and Bauchi, this method of selection was successful; no selected village had to be replaced for being atypical or for noncooperation as the studies proceeded, even though, in the Zaria case, the original estimate of one or two years for a long-term relationship extended into eleven years.

The three villages in the Zaria area selected for the descriptive basic studies of farming households and their farming operations were (Table 5.1 and Map 5.1):

1. Hanwa, on the outskirts of Zaria.
2. Doka, about 40 km from Zaria along the two-lane paved highway connecting Zaria and Kano.
3. Dan Mahawayi, about 32 km from Zaria, but reached primarily via dirt roads and tracks, the last 11 km of which are easily motorable only in the dry season.

Study Sequence

The sequence of fieldwork which ultimately led to the articulation and appreciation of the farming systems approach to research began with a straightforward attempt to try to understand what farmers were doing. Attention initially was focussed on production aspects. The agricultural economists heading the teams adopted a farm-management mode of analysis, measuring the use of and

MAP 5.1
Villages Included in the Zaria Studies

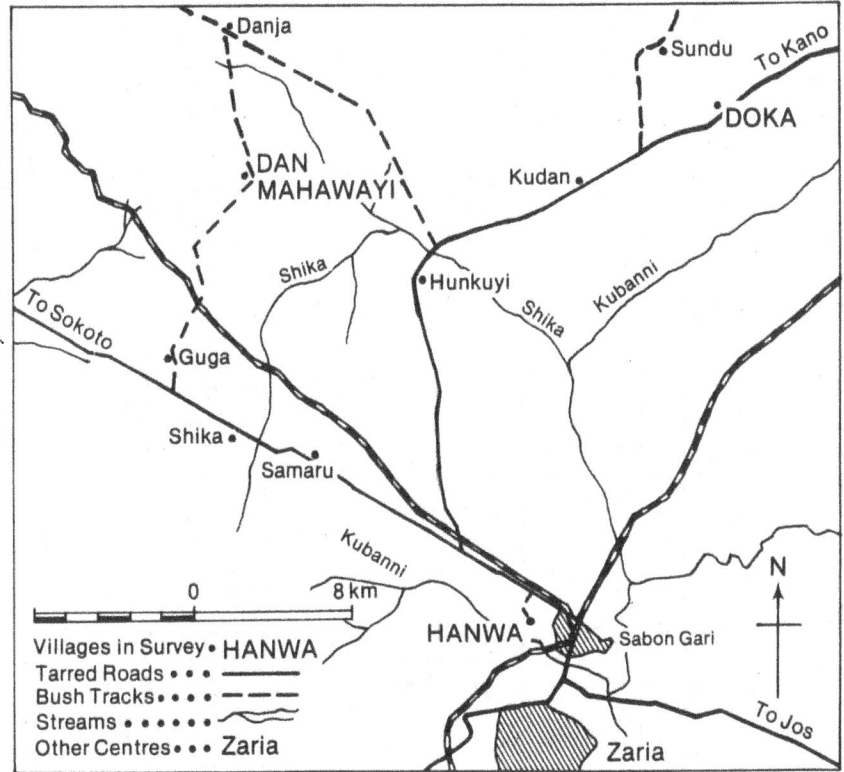

returns to the primary factors of production--land, labor, and
capital--and employing a cost-accounting approach to the collection
of farm-budget data. Other team members looked at personal and
community networks, farmers' organizations, and the adoption rates
for various farming practices, both traditional and modern.

In analyzing the first set of quantitative data (mainly
relating to production activities), it became clear that the
information set needed to understand the farming household and to
evaluate the constraints of farm decision-making was incomplete.
Detailed data on marketing and storage operations and on incomes and
seasonal consumption were obvious deficiencies. Thus, two
separately designed studies were initiated in 1970-72 in the same
Zaria villages to survey the same sample households surveyed in
1966-67. To evaluate in-kind incomes more precisely than had been
possible in the farm-management study, we asked members of the
sample households in each village to participate in a survey of
expenditures (usually referred to as the consumption study). As
food items had to be weighed to impute cash values to them, the

TABLE 5.1
Major Studies Undertaken and Characteristics of the Zaria Villages[a]

Variable Specification	Dan Mahawayi	Doka	Hanwa		Average of the Three Villages[b]
			Non-cattle Owners	Cattle Owners	
Location	11° 19'N 7° 35'N	11° 22'N 7° 47'E	11° 3'N 7° 43'E		
Ease of accessibility to Zaria	Poor	Good	Very Good		
Population density per sq km (1965)	32	153	274		153
Number of households in villages (1966)[c]	109	153	88		117
Number of households in each study:					
Production (Jan. 1966-April 1967)	42	44	20	18	41
Marketing (Sept. 1970-March 1972)	18	18	11	7	18
Consumption (April 1970-July 1971)	41	43	19	18	40
Occupations of women (April 1970-May 1973)	47	d	35		27

[a] The list includes only the main studies, which provided the empirical data for this chapter.
[b] Unless otherwise stated, the data in this column in other tables in this chapter represent the average from weighting each village equally. In deriving the average for Hanwa, cattle owners and non-cattle owners were weighted equally.
[c] Dates of the studies are given in parentheses. The sociological studies, undertaken from 1966 to 1976, are not listed because many different samples were used.
[d] Some data were collected from this village but not analyzed.

expenditure/consumption study also permitted a fairly detailed description of consumption from a nutritional perspective. About half of the sample households in the expenditure survey were included in the concurrently conducted marketing/storage survey. In that study extensive data were also collected on credit transactions and the grain trade.

Subsequent analysis of the data set confirmed the apparent significance of off-farm employment and cash incomes, particularly by women. That led to the design of another village study, focussed primarily on women's income-earning occupations and carried out intermittently between 1970 and 1973.

All parts of the Zaria survey were complemented by, although not integrated with, a series of sociologically oriented studies in the same villages throughout the 1966-76 period. In retrospect, it would have been better to have carried out the quantitative basic studies and the softer sociological studies simultaneously so that the research could have been integrated more fully. But lack of staff at the beginning, the evolving nature of the social science research program at IAR, increasing teaching commitments, and, of course, the lack of the advantages of hindsight, account for the remaining presence of some frustrating gaps in our understanding and analysis of Zaria farmers' decision-making behavior.

Sample Design and Survey Methods

The sample design for the farm-management study was fundamental in defining the relationship between specific farmers--that is, farming families or households--and various researchers over the year. Whereas the villages were purposively selected, only the village heads' households were similarly chosen. It was essential to include their households in the sample for reasons of protocol, status, information, and good community relations. The remaining 121 households in the production or farm-management study, however, were randomly selected after complete enumeration of each village. There was no stratification at the sampling stage; approximately 35 percent of all households were included in the sample, although the sampling fraction varied from village to village.[2] Total sample size was determined by assessing potential enumerator workloads.

Whenever possible, the households included in the farm-management survey were also included in the household expenditure/consumption survey. Because fieldwork for the marketing/storage study began shortly after the consumption study was started, the marketing sample was based on the consumption-study sample. It was believed that the detail of the marketing/storage/ credit/trade questionnaires would require more enumerator time per interview than the consumption/expenditure survey. Thus, only half the sample households were randomly selected for participation in the marketing survey (Table 5.1).

The women's occupation survey originally was designed to include all women in the expenditure/consumption survey households, but that proved unworkable for various reasons. There were conceptual problems in designing a survey instrument, a very large number of women were in the sample, and the combination placed too many demands on enumerators' workloads.

Enumerators assisted with all fieldwork. Male enumerators lived in each of the study villages from 1967-73; female enumerators were in residence only in 1970-71. Because no employable residents in the study villages were literate, all enumerators were Hausa-speaking "strangers."[3] Each enumerator was supervised directly by the researcher in charge of a particular study, normally through weekly or twice-weekly, day-long visits at a minimum. Other visits, to collect forms, deliver paychecks, and resolve problems, were made to the villages by administrative assistants, who also filled in when enumerators were ill or on leave.

In each survey a series of structured questionnaires and forms was used to record quantitative data consistently throughout an entire year. Open-ended interviews with various survey participants, and often nonparticipants, as well as extensive participant-observation were also employed on occasion to address various issues. All studies included some direct measurement--for example, of fields, yields, plant spacing, and food consumption. The underlying theme of all data-collection modes and supervisory techniques was to minimize measurement, or nonsampling, error in quantitative estimates. In addition to formal and informal interview settings to focus on the research topics at hand, much time was spent in villages engaging in conversations and discussions. Some of these conversations were more relevant than we realized at the time.

It was apparent at the outset that, to achieve the desired degree of quantitative accuracy, it would be necessary to reduce memory loss through frequent interviewing and verbal reports, confirmed visually through direct measurement. Though no sophisticated tests were conducted to come up with the most cost-effective survey intervals, simple checks on data quality showed that, for certain variables, reports had to be solicited daily. Still other variables were reasonably accurate when reported weeks or months after the event; others had to be measured directly because any verbal report was hopeless.[4]

For both farm-management and marketing studies, each sample household was interviewed twice weekly throughout the survey year. In the case of the expenditure/consumption study, each sample household was interviewed daily for two non-consecutive weeks during the survey year, and each time many food items were weighed. The women's occupations study was a test of patience and conceptual revision; developing an effective survey instrument took more than a year and dozens of interviews.[5]

Chapter Outline

We now turn to a discussion of the results of the various studies. Our major concern at the time of the studies, as we have already noted, was to explore broadly the ways in which farming households made decisions on various aspects of their farming systems. According to the mandate of the Rural Economy Research Unit, we were to feed back that information to other researchers at the Institute for Agricultural Research to help them understand why Zaria farmers--most of whom were relatively untouched by current agricultural recommendations--were so reluctant to adopt modern

agricultural technology.
     Though the feedback process was somewhat unsystematic at the time, we here take advantage of the perspective offered by distance in both space and time to formulate a reasonably accurate idea of farming systems in Zaria. In one way, that could seem to be a sterile recapitulation of facts already out of date. On the other hand, we see the Zaria research experience and results as a means of illustrating empirically: first, the variety of issues likely to confront any research effort in which a farming systems approach to research is used, and, second, the variety of ways in which analysis of those issues can affect the understanding gained.
     Perhaps the most fundamental issue that faces teams involved in the farming systems approach to research (FSAR) is that each farming household operates its own unique farming system. At times it may seem impossible to generalize except by calculating means and averages. FSAR researchers must, however, attempt to identify constraints that are critically important in keeping each farming household from achieving its own goals or from implementing change, as well as to identify those shared by a number of households. Agricultural research, to be economically feasible, must lead to conclusions relevant to a significant number of farming households; the expense of providing one extension worker for one farmer would be too great for even the wealthiest society. Finally, agricultural strategies and policies are generally fairly blunt instruments for change. If FSAR is to result in recommendations for policy changes or for revising agricultural strategies, a substantial number of farmers must be involved.
     How, then, can one identify this substantial number of real, not average, farmers who share constraints similar enough to be addressed by research workers, extension agents, and policy makers? Village surveys and farm-management studies are standard responses; both, though time-consuming and expensive, permit one to obtain, as we shall show, greater in-depth understanding of complex farming systems than can be achieved from rapidly analyzing pre-selected key indicators such as size of farm and gross production.
     We use the farming systems framework presented in Chapter 2 as the organizational framework here. Since the technical element and several exogenous factors were discussed in general terms in the preceding two chapters, and, by definition, they are common to nearly all farming systems in the area, the major emphasis in this chapter is on the endogenous factors. After discussing the endogenous elements of the farm household's decision-making process, we briefly analyze the resulting crop, livestock, and off-farm processes. We conclude this chapter with a discussion on the implications arising from the analysis.

THE TECHNICAL ELEMENT

Physical Factors

     The landscape consists of gently undulating plain 600 to 900m above sea level. Inselbergs rise above the plain in some places, but broad valleys are common. The leached, ferruginous tropical soils characteristic of the Northern Guinea ecological zone

FIGURE 5.1
Rainfall and Dates of Planting and Harvesting of Major Crops, Zaria Area[a]

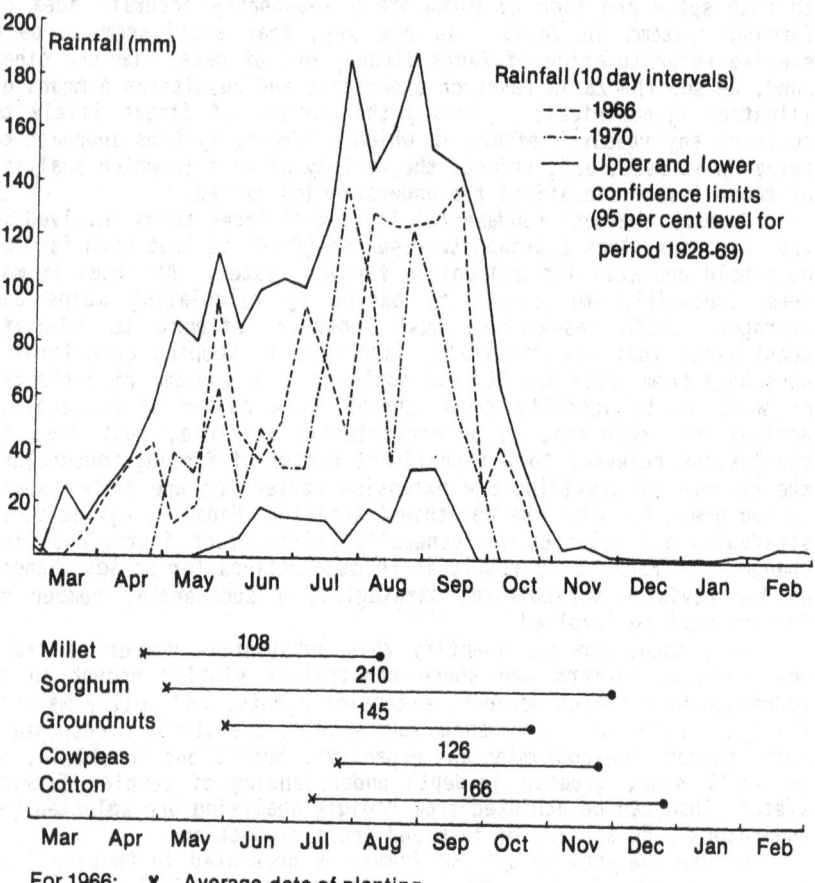

For 1966: **x** Average date of planting
  • Average date of harvesting
  108 Average number of days crop is in ground

[a]Rainfall data are given in 10 day intervals while the crop data refer to 1966-67.

naturally supported savanna woodland, but because of human activity woodland has been largely replaced by parkland.

The critical technical variables with which all Zaria farmers must cope involve land and water. The amount and distribution of rainfall are critical factors in determining crop yields, which justify the labor required to make crops grow and to protect them from pests and predators. Labor is the primary investment that a farmer makes in the land; hoeing, ridging, weeding, harvesting, and threshing are, in the Zaria area as in much of the Nigerian savanna, tasks that involve only human effort. The relationship, then, between the farmers and the land they use is an intense one, with the degree of intensity being heavily influenced by the amount and distribution of rainfall. The average annual rainfall in the Zaria area is 1,105 mm, which falls mostly during the period from April to October. A severe water deficit exists during the dry season from November to March. The figures in Figure 5.1 for the meteorological station nearest the survey villages show that the 1966-67 crop year was within the 95 percent confidence limits of rainfall averages for the period 1928-69 and somewhat on the high side of normal (that is, 1,332 mm). The marketing study was in the year after the production year of 1970-71, when the rainfall was slightly less than normal (that is, 948 mm) and was in fact the precursor of the Sahelian drought of the early 1970s.

The degree to which farmers rely on rainfall to some extent depends on the type of land they farm. Gona, or rainfed upland, fields support the seasonal production of grains, legumes, cotton, and root crops, whereas low-lying bottom lands, or fadama fields, are close enough to the watertable to permit year-round cultivation of such crops as vegetables and sugarcane. Though the cultivation of low-lying fadama fields would appear to give farmers some insurance against the effects of poor rainfall, serious and long-term reduction in rains obviously will affect the watertable level in fadama areas as well. During our study, no farm households undertook any method of pumping or water control. Furthermore, all farm households had access to the use of gona for cultivation, but not all had access to the use of fadama. Doka was selected as a study village precisely because it is well-endowed with fadama acreage; all but one Doka farmer in the sample reported cultivating fadama fields. Nevertheless, reliance on rain and upland soils was a predominant feature of all farming systems.

Biological Factors

Gona land was thus very influential in determining the pattern of crops grown. The close correlation between the rainfall distribution and the growth cycles of the main crops grown is illustrated in Figure 5.1. The figure illustrates the role of early millet (gero) as an early-in early-out crop.

Millet was harvested at the height of the rains. Though that may make for some drying problems, the availability of one staple three months before the other was harvested did ensure that a reduced yield or failure of one harvest because of maldistribution or insufficiency of rainfall would not affect the growing cycle of the other staple in the same way.[6]

Forage was not explicitly grown or harvested to feed livestock in the dry season, so the condition of the livestock was also a function of rainfall. Because of lack of feed and its poorer quality during the dry season, cattle owners herded their cattle considerable distances from the village in search of food. Also, in contrast to the rainy season, when goats and sheep were restrained to prevent damage to growing crops, they were left to find their own food during the dry season. It is therefore not surprising that livestock, particularly cattle, lost condition during the dry season.

THE EXOGENOUS INFLUENCES

Just as the physical and biological facts of life define the potential crop or livestock yields of farming systems, exogenous factors, which are socio-economic in nature, constrain the range of choices that any farm operator can make--with respect to crops, occupations, expenditures, and even livestock ownership. Trying to understand just which factors are relevant, however, is easier said than done. Contrast two households: each with ten members, headed by men in their thirties, each considered large farmers by village scales, each selling or giving away approximately 750 kg of sorghum and millet in 1970-71. One household operated in gandu, supplemented family labor with substantial amounts of hired labor, but pursued no off-farm occupations regularly. The other farm was managed by the household head, who supplemented his own labor with only a minimal amount of hired labor and reportedly traded in only one seasonal crop, sugarcane. Both reported roughly equivalent grain sales and gifts in 1970-71, but for one household such sales and gifts represented 60 percent of total grain production; for the other, only 28 percent. One household reported consumption levels well below calorie requirements; the other exceeded calorie requirements by more than 10 percent, on the average.

Trying to understand what makes for such contrasts in performance and in decision-making behavior is the challenge of analysts; the challenge of development practitioners is to try to change those exogenous factors that keep rural households from succeeding, either in their own or in societal terms. The farming systems approach to research implies meeting both challenges.

Some of the many exogenous factors that influence farming decisions have already been described, particularly in Chapter 4. Community structures, norms, and beliefs as well as external institutions--both public and private--place certain bounds on all farming families' decisions. However, differences in population density and accessibility to the outside world are partially responsible for the substantial variation within those parameters. In addition to directly influencing the farming systems that farming families can adopt, these factors are also contributing indirectly to some of the changes that are occurring in the farming systems practiced by farming families.[7]

Population density and demand for land determine the area of land available per farm and the ability to fallow; status within the community structure, income, and the degree of commercialization of land determine a farming family's ability to claim more. The availability of nonfamily labor constrains farming households' abilities to consider employing it; cash availability further qualifies the ability to hire additional labor. The access to markets determines both selling and purchasing behavior for both products and inputs. The prevalence of house trade and of independent rural assemblers moderates this accessibility for certain commodities. Prices are largely determined outside the village; farmers thus have limited bargaining power regarding price and find their purchasing power influenced by exogenous factors, modified, of course, by the extent of their ability to be largely self sufficient in food production. We look here at the impact of each of these influences individually as they shape farming systems and then at the combined effects of all factors as a set.

Land Supply and Demand

The average amount of land available for a farming household is clearly related to the density of the area's population. As population density increased with increasing access to Zaria, households in Hanwa, the village closest to Zaria, had the smallest farms on the average (Table 5.2). And where farm land was in shortest supply, holdings per household also tended to be more equitably distributed. Two reasons could be posited for that:

1. The village head's traditional distributive power over holdings was exercised to ensure equity of access to all who wanted or needed land to farm, especially where the land resource was particularly scarce.
2. Competition among individual farmers for farms of a viable size led to a relatively narrow range in sizes.

There was some evidence for rejecting the first hypothesis. The village heads did not exercise their power to increase the equity of holdings. If anything, the fact that Fulani cattle-owners' farms in Hanwa were nearly twice as large as those farms held by people not owning cattle is an argument for ethnic bias in land availability. Although that cannot be substantiated by documentary evidence, the difference in farm holdings was great enough to justify treating cattle owners (largely Fulani) and non-cattle owners (largely Hausa) as distinct agricultural subgroups for most analytical purposes. Competition, however, seemed to be a more salient hypothesis for the somewhat greater degree of equity in size of holdings in the more densely populated areas. The evidence was circumstantial, but persuasive.

Fully 74 percent of farmland in Hanwa had been acquired for either temporary or permanent use by renting, by borrowing, or through outright purchase of usufructuary rights. The opportunity cost of leaving land fallow in such areas was relatively high; farming households in Hanwa were thus encouraged to surrender their usufructuary rights if they did not farm the land themselves.[8] In Doka and Dan Mahawayi, by contrast, most farmland was inherited, so fallow ratios in 1966-67 were much higher (Table 5.2).

As Zaria encroaches further onto the outlying fields of Hanwa, it is likely that increasing pressures will be placed on the land-tenure system, with land "purchases" becoming even more common. For example, in 1973, one farmer in particular was known to be purchasing land for his sons and building them individual compounds along the village's main path. He was doing that, he said, so his family would be ready to take advantage of the growth and increased land values, to say nothing of paved streets, which he believed would soon mark Hanwa's entry into modern times.

One other effect of the interaction between the supply of and

TABLE 5.2
Land and Labor Availability, Zaria Villages, 1966-67

Variable Specification	Dan Mahawayi	Doka	Hanwa		Average of the Three Villages
			Non-cattle Owners	Cattle Owners	
Land:					
Farm size:					
Average (ha)[a]	4.8(21.2)	4.0(26.8)	2.2(2.5)	3.7(2.4)	3.9(16.8)
Gini coefficient[b]	0.55(0.49)	0.33(0.30)	0.34(0.34)		0.41(0.38)
Composition of farm (ha):[a]					
Gona	4.4(19.2)	3.5(29.0)	1.9(2.7)	3.6(2.4)	3.5(16.9)
Fadama	0.4(42.1)	0.5(12.4)	0.3(1.5)	0.1(0.0)	0.4(18.4)
Land inherited (%)	77.0	91.8	30.0	20.0	65.0
Labor:					
Household (av. nos.):					
Residents	6.8	8.0	10.3	11.6	8.6[c]
Consumer units[d]	4.9	5.8	7.5	8.2	6.2
Male adults	1.7	2.0	2.9	2.8	2.2
Households of gandu type (%)	42.9	40.9	45.0	83.0	49.3
Ratios:					
Hectares per:					
Resident	0.8	0.5	0.2	0.3	0.5
Consumer unit	1.0	0.7	0.3	0.4	0.7

[a]Figures in parentheses indicate the percentage of land fallowed.
[b]Figures in parentheses represent the gini coefficients for cultivated land.
[c]The 20:80 percentile points of household size were 5:12.
[d]The consumer units were based on dietary requirements suggested by the FAO (1967).

demand for land is fragmentation of farms. Fragmentation has been steadily increasing, in part due to the natural process of inheritance among sons, but accelerated perhaps by the reduction in gandu farms and increase in iyali farms.[9] In 1966-67 the average Zaria farm of almost four hectares consisted of six or more different fields, a type of fragmentation often cited as a potential block to progress because it does not permit mechanization of field work. But as long as hand labor provides the main source of power, fragmentation provides some distinct advantages, which revolve around a notion of equitability different from the one based on size --for example, in distributing land of different soil types, minimizing the effects of microvariations in rainfall (particularly at the beginning and the end of the rainy season when such variations can be critical), and distributing among farming households the inconvenience of distant field locations.

Community Status and the Degree of Commercialization

The impact of a household's status in the community and the effects of commercial opportunities on the market for land have already been noted. But community status of households and market opportunities influence other aspects of the farming systems as well. A dramatic illustration was found in Hanwa, where the village population was about equally divided between Fulani and Hausa. Those who were Fulani tended to own cattle and to operate larger farms; Hausa households did not own cattle, had smaller farms, and the men in these households tended to pursue off-farm occupations more regularly. Ownership of cattle, as opposed to nonownership, moreover, implied that the difference observed between Fulani and Hausa male work behavior also extended to women. Fulani women in Nigeria traditionally control the marketing of milk and butter. With the availability of good retail markets in Zaria for those products, combined with the proximity of the village to the markets, it was not surprising to observe, during our study, that women from Fulani households tended to spend a good part of each day outside their compounds, moving freely in public. The ready-to-eat food markets in Zaria also offered such lucrative commercial opportunities that women in Hanwa, Hausa as well as Fulani, tended to engage more steadily in entrepreneurial enterprises involving food products than did women in Doka or Dan Mahawayi. Dan Mahawayi included a mix of Hausa and Fulani households as well, but ownership of cattle was rare and the commercial opportunities for milk did not approach those in Hanwa; hence, settled Fulani women in Dan Mahawayi behaved much like Hausa women there--observing purdah and pursuing the more limited range of economic activities for which there was a local market.

In both Doka and Dan Mahawayi, therefore, a household's community status was less linked with ethnic origin and was more readily related to income and/or political status--which were also highly correlated.

The Availability of Non-Family Labor

None of the households in the survey villages had mechanical equipment for use in cultivating the fields. When reliance is on hand labor, the maximum size of any household production unit is determined in part by the availability of people to use hoes, hand plows, and other hand tools. A class of landless agricultural laborers did not yet exist in the Zaria area. Although, as we shall discuss later, there were variations in the use of hired labor, in most households relatively little such labor was used in 1966-67. In fact an average of only about 18 percent of the total labor input on the family farm originated from non-family sources (Table 5.3). Nevertheless, though we made no direct tally of the number of people looking for work, the farm-management study provided evidence that a market for hired farm labor did exist. Approximately 330 hours of nonfamily labor were purchased by the average farming household. Farmers in Hanwa availed themselves of the largest quantity--475 man-hours per farm per year. Interestingly, however, farming families in Dan Mahawayi employed nearly as much nonfamily labor as did those in Hanwa in absolute terms, and more in relative terms: 29 percent of the total farm labor in Dan Mahawayi was hired, compared with 21 percent in Hanwa. Supply in Dan Mahawayi could be posited to have been a function of the lack of other employment opportunities, whereas supply in Hanwa could have been more of a response to demand. Wage-rate information did not bear out that supposition, however; the average wage in Hanwa was actually lower than in Dan Mahawayi.[10] In fact marginal-productivity information would tend to support a hypothesis that demand for nonfamily labor in Dan Mahawayi encouraged more people to offer themselves for hire. In Hanwa, on the other hand, the amount of nonfamily labor available was ample and could be used to the point at which the marginal value product was closer to zero.[11]

Access to Markets: Selling and Purchasing Behavior

In earlier discussion (Chapter 2), we emphasized that one of the necessary conditions for the adoption of improved technologies by farmers is the adequacy of external institutions or support systems on both the input and output side.

On the input side, adequacy can be interpreted as the presence of the required improved inputs--such as inorganic fertilizer, improved seed, and seed dressing; an extension agent to provide instruction on their efficient use; and possibly an institutional credit program to facilitate their purchase. During the time of our studies in the Zaria villages, no institutional credit programs operated, no extension agents lived or worked in the villages, and no improved seed apart from cotton was officially distributed. Other improved inputs, such as seed dressing and inorganic fertilizer, were either unavailable or available in such limited quantities that they were being sold at inflated prices. It is therefore not surprising that little in the way of improved inputs was used by farmers in the Zaria villages during the period of our studies.

On the output side, we already have noted the differential access to markets as being of some importance in influencing women's work patterns and household landholding patterns. But, as might be expected from our earlier discussion (Chapter 4), markets also play

TABLE 5.3
Work on the Family Farm, Zaria Villages, 1966-67[a]

Variable Specification in Terms of an Average Household	Village				
	Dan Mahawayi	Doka	Hanwa		Average of the Three Villages
			Non-cattle Owners	Cattle Owners	
Annual man-hours on the family farm[b]	1516	1634	2109	2405	1803
Source of farm labor as a percent of total man-hours:					
Family: Male adults	62.9	82.2	76.5	65.4	72.3
Female adults	0.1	0.3	0.7	0.6	0.3
Older children	8.3	10.9	9.9	5.0	8.9
Hired: Kwadago	14.3	1.4	6.2	13.6	8.5
Jinga	14.1	4.6	6.1	11.1	9.1
Gaya	0.3	0.0	0.6	4.3	0.9

[a]The figures in this table exclude time spent travelling to and from fields.
[b]Conversion to man-hour terms is given in note 12.

more pervasive roles in household decision-making behavior and in determining the effect of those decisions on income.

Farmers in Hanwa had direct access to the regional market in Zaria, an important factor in the grain and vegetable trades. Average grain prices in Hanwa were 22 percent higher than in Dan Mahawayi because there were fewer people in the marketing chain and transport costs were 10 to 15 percent lower (as a percentage of the margin). In addition to farmers receiving lower prices for their produce in more distant villages, such as Doka and Dan Mahawayi, their families also had to pay more for consumer goods imported from outside the region. The price of rice, for example, was 18 percent lower in Hanwa than in Dan Mahawayi in the dry season and 3 percent lower in the wet season (Simmons 1976c). Clearly, if the farming households wanted to market their own products in the urban market, thus performing some of the marketing functions of the marketing intermediaries, they could receive a higher price. After allowing for transport costs, farming families could receive that return normally accruing to the intermediaries. Thus, the decision of farming households to market their own product would depend on two variables: transport cost and the opportunity cost of their own time. Generally, given the nature of grain marketings, which were often in small quantities, most farmers chose to market their grain in the local market through intermediaries. In the case of cash crops, when larger quantities could be marketed at one time, farmers in Doka did often market their own crops (e.g., groundnuts, peppers, onions). Farmers in Dan Mahawayi did that less frequently because of the difficulty of obtaining truck transportation. An important factor influencing marketing costs and thus marketing efficiency of the Doka farmers had to do with their ability to assume the functions of the marketing intermediaries and to market their own product if they deemed the marketing margin of intermediaries excessive. Before that could be possible in Dan Mahawayi, roads would have to be improved so as to decrease transport charges and improve accessibility to markets.

Exogenous Influences: a Summary

We have briefly described some of the effects some exogenous influences can have on shaping a household's decisions. We confined our discussion largely to village-level contrasts because virtually all households in a given village face the same exogenous constraints: access to land, markets and hired labor, prices and socio-cultural parameters. The marked exception is the status ascribed to various households in the community. Though certain aspects of this status are exogenous in character, especially ethnic origin, other aspects are closely related to the success or character of individual households and household heads. Thus, status will also be considered in that endogenous context in the next section.

As we examine (in the next section) the various responses of farming households to the exogenous factors, we can begin to see how individual farming families work within them and achieve various levels of productivity. It is obvious that exogenous influences can contribute in many ways to inter-village differentiation. They may, for example, be partially responsible for differences between villages in terms of the productivity of gona and fadama, the productivity of hired and family labor, and the composition of crop enterprises and off-farm employment. Because of that, the following analysis of endogenous influences is complicated by influences that are exogenous in nature. Therefore, on occasion, to clarify relationships in correlation and regression analysis, we try to isolate the exogenous influences through the use of variables denoting specific villages.

THE ENDOGENOUS INFLUENCES

It is the endogenous conditions of the households themselves that differentiate among them and the decisions they make--how they develop and apply their skills and improve the productivity of their resources, how they identify themselves with and contribute to the society in which they live, how they spend their inheritances and build their wealth, and how they satisfy their needs for food. In this section, we look at the households surveyed to gain a better understanding of the operation of the decision-making process within them--how households facing similar exogenous conditions match their resources to their objectives by mobilizing those characteristics that are part of the household itself--people, skills, ideas.

The number of people in a household, the age and sex composition of that group, and its organization in gandu or iyali are perhaps the most fundamental differentiating factors in farming behavior at the level of the production unit. By examining the allocation of the household labor force to the various tasks which together determine household productivity, we shall show that the allocation is to some extent conditioned by such households' participation as members of communities. That is particularly so with regard to socially accepted definitions of women's work, as well as to such inseparable exogenous influences as hired labor supply and work opportunities. Despite those conditions, however, households in our study exhibited considerable variation in work-allocation decisions. We can treat investment decisions only briefly, in that quantitative information on them is sparse, even though we recognize that the acquisition of new resources, both technical and human, may be critical for a household's long-term achievement of its goal(s).

Household Size, Composition, and Organization

Household size may be viewed as the result of farm and household decisions or it may be viewed in the opposite light--as the cause of them. Rather than argue for either the chicken or the egg, we simply note here that household size and composition must be taken into account by farming systems analysts, just as the heads of households take them into account in their own planning.

At the start of our farm-management survey, the average Zaria household included 8.6 members: 2.2 male adults including the household head; 2.6 female adults including wives, mothers, and others; and 3.8 dependent children. Just about half of the Zaria households were organized in gandu (Table 5.2) with their average size being 10.9 individuals, compared with 6.3 persons for an iyali. The iyali households could be further disaggregated; at the time of the household expenditure study, 40 percent included sons who were not yet married but were old enough to work nearly full-time on farming operations, and 60 percent either had no sons of working age or no sons at all. The head of an iyali household was estimated to be on the average, 47 years old; in contrast, a gandu head was, on the average, 51 years old.

Such broad averages are useful in sketching a quick picture, but more details must be added if an analyst wishes to understand relationships among household demographic variables, farming systems, and farm productivity. That the needs of cattle owners in the villages we observed differed from those of farming families owning no cattle, for example, was reflected in the gandu organization of their households and hence in the relatively greater sizes of cattle-owning households. Eighty-three percent of the cattle owners were organized in gandu, and the average cattle-owning household had one more member than did the average household in the same village owning no cattle (Table 5.2).

Cross-sectional analysis of the Zaria village data showed in general that the larger the household, the larger the area the household would be expected to cultivate (Table 5.4), given, of course, the village-level constraints on land availability already noted.[13] Causality, however, was not indicated by the correlation. The relationship can be explained in two ways: first, with no mechanization, more laborers meant more work capacity; or second, with more mouths to feed, more space was needed to grow sufficient food. That is, assuming size of the farm-production unit was roughly related to the scale of production, the size of the household "eating from one pot" was also related to the scale of consumption. Indeed, in one regression analysis, household size appeared to have more effect on total food consumption than did income; a doubling in household size was estimated to cause a more than 40 percent increase in all food expenditures, and a more than double increase in sorghum consumption (Simmons 1976c).

But were the individuals in bigger households generally richer and better-off? As might be expected, the situation was more complex than it at first might seem. There did not appear to be any economies of scale, either in production or in consumption, but there did appear to be some relation: first, between total household income and household size; second, between household farm income and mode of organization, with gandaye in general being better off than iyalai; and third, between household size and per capita level of calorie intake. Though the first two relationships were positive, indicating group success, the consumption relationship (which begins to address individual welfare) was negative (Table 5.4). Indeed, the significance of the first two relationships changed or disappeared when income was expressed in per-capita terms. That underscores an analytical caution for farming systems researchers: it is important to separate group observations from those of individuals.

TABLE 5.4
Partial Correlation Coefficients Between Level of Well-Being and Households, Zaria Villages[a]

First Variable	Second Variable		
	Name	When Expressed in	
		Total Terms	Per Capita Terms
Household size	Cultivated hectares	0.5864*	-0.0574
Household size	Disposable income	0.4556*	-0.2821*
Household organization[b]	Disposable income	0.4613*	-0.0914
Household size	Calorie intake		-0.3930*

[a]Second order partial correlation coefficients were calculated with two variables controlling for village location. For the first three coefficients, 1966-67 farm management study data were used (sample size (N) = 124 households), whereas the last one involved using the 1970-71 consumption study data (N = 109 households).
[b]Iyalai households were weighted as 1; gandaye households as 2.
*Significantly different from zero at the 5 percent level.

The point of the life cycle at which a household was when surveyed is also likely to be an important classification variable, but in such Hausa societies, life-cycle status was difficult to define. For example, young men just beginning their married lives as junior partners in gandu were likely to behave quite differently from those who became heads of households in a small and separate iyali where they farmed property inherited upon the death of a father or those household heads in a small and separate iyali where they were trying to acquire land and other assets by their own labor. Similarly, vigorous heads of gandu still acquiring wives and working sons were likely to make significantly different farming decisions from heads of gandu whose sons were leaving. Our data were not adequate to handle this variable well.[14]

The Household Labor Force and Household Productivity

In a semi-subsistence--or semi-commercial, depending on one's viewpoint--rural economy, such as that of the Zaria villages, it can be assumed that the work effort of household members would be directly related to household productivity. Possession of fixed assets of the level and type needed to generate significant amounts of unearned income was generally not characteristic of village households, with the possible exception of a few large landholders or titled position-holders. In this section, therefore, we focus on the household labor force in the Zaria villages and the factors that determined the amount of time members of that labor force were likely to devote to work. Among these factors were:

1. The responsibilities that various members bore toward providing for the household welfare.
2. The physical ability they had to work, which is a function of health and nutrition levels as well as of age and sometimes sex.
3. The resource base owned by the household and/or its members, that is, land, cash, and cattle, and the technical productivity of that base (e.g., fadama compared with gona).
4. The farm-production demands that were largely a function of season and crop choice.
5. Opportunities for nonfarm work, in which sex, status/asset endowment, and location were important.
6. Work incentives, which, in turn, were based on the potential returns to work and to the need for the production of that work--food primarily, but also assets such as wives, land, and "security."

Responsibilities for work. In Zaria households, the division of responsibilities for various domestic tasks and for various contributions to household welfare was relatively clearcut and fairly widely followed by all households within a village. The responsibilities obviously change somewhat over time, but during the decade or so of our research in Doka, Dan Mahawayi, and Hanwa, they appeared to change remarkably little.

As observed in our surveys, adult men were responsible for providing food for the major meal of the day, for shelter, and for clothing at the time of the major Islamic celebrations, particularly Eid el Fitr. The head of the Zaria household, of course, assumed primary responsibility among all household males, but the actual activities involved were commonly carried out by the subsidiary males in the household. Most households we observed, for example, had rules for daily allocation of the grain stored in the compound's rumbus; measuring withdrawals of grains produced on the farm was important because it permitted close monitoring of the household's ability to feed itself throughout the year. Removing grain from the granaries was exclusively a male task in all villages; though the male responsible usually was the head of the household, this activity often was delegated to a brother or son. The bundles of millet and sorghum generally were taken out one or two at a time every few days and given to the women for processing and preparation. In some savanna villages women help with the production of crops, but in the Zaria villages their responsibilities in food preparation began when the grain was removed from the granary. Unlike women in many other parts of Nigeria and the West African savanna, Zaria women played no independent roles in farming and left the production, storage, and withdrawal from storage of the major foodgrains to men. Men also tended to assume major responsibility for shopping for food in markets, for carrying water, if necessary, and for their own laundry. Any other work activities in which men wished to engage, for household benefit, were up to their individual initiative and discretion.

Adult Zaria women were responsible for cooking and other food processing and for bearing and caring for children. Because of the customs of seclusion and the methods of house construction, compounds generally included several adult women. Women's responsibilities for household work were thus often shared on a rotational basis; that is, each woman performed a certain task for a fixed number of days. Women's allocation of available work time to nondomestic, nonchildcare tasks rested on their personal decisions. Most pursued independent money-earning activities, using the proceeds to purchase supplemental food for themselves and their children, to give gifts to friends, and to provide for such personal needs as soap, clothing, and cosmetics. They made great efforts to save some of the profits from their independent enterprises; savings were most often held in the form of enamel and brass bowls, cloths, and sometimes mats, beds, perfumes, and other items purchased in anticipation of their daughters' marriages. It should be noted that women who divorced their husbands, or were divorced by them, bore no further responsibility to their children who had been weaned. Often, however, women continued to prepare for daughters' marriages and also gave small gifts to their children who remained behind in the father's compound.

Children under the age of about fourteen had responsibilities such as performing errands for adults, caring for children in the household, and doing certain farm or cattle-herding tasks at the request of their parents. Children played absolutely critical roles of communication with the outside world for their secluded mothers, and generally helpful roles for their fathers and older brothers (Schildkrout, in press).

Young adults were those children who had begun to bear some responsibility for their own welfare. Girls, because they were married so young, started to collect items for their marriages at a very early age, generally by working for their mothers, although sometimes on their own account. After marriage, however, young women becoming adults were not expected to assume independent roles too fast, but were instead expected to take on many of the domestic responsibilities of their mothers-in-law. Most young women, before the bearing of their first child, did not, for example, engage in an occupation, but merely did the household cooking and cleaning and assisted the mothers-in-law in their money-earning activities. On the other hand, boys who expected to inherit land from their fathers were expected to contribute to their own marriage payments; fathers often helped out. Boys did, however, have an incentive to begin to engage in independent work tasks on their own as soon as they were able, but the fact that young men tended to get married at much later ages than girls did may have reflected the constraints on abilities of boys to accumulate financial resources in adequate amounts on their own.

Where domestic economic roles are so clearly defined, it is interesting to note what happened when the person designated to take on a particular responsibility was not available, or when that person for some reason could not bear that responsibility. If there were no adult male to head a household, for example, did women take over the farming that was the basis of the household's survival? If there were no children, did women break seclusion to do errands themselves? If a farmer had too little land to produce an adequate amount of food for his family, did his wife help with the cash proceeds of her occupation?

The answers to such questions are perhaps the key to the flexibility or rigidity of a particular socio-cultural situation and to its adaptive behavior. Not all such hypothetical questions can be treated here, but two ways in which shifts in responsibilities engendered by change and modernization were handled in Zaria illustrate the importance of gaining such understanding.

1. Men were nearly always the heads of rural households in the Zaria area; few women retained the rights to the land they inherited. When a male household head died, one of two possibilities usually occurred: first, the household broke up and the widowed wives went back to their own fathers or (if the fathers were dead) to their brothers' or sons' homes; or, second, a new male head was designated, as when a brother-brother gandu succeeded a father-son gandu--in which case, the widowed wives might continue residing in the compound then headed by their sons. Rarely did rural women become heads of households in Zaria; when they did they were rarely able to support themselves through farming or their traditional occupations (Longhurst 1980). They thus subsisted on charity from others and were anomalies in the rural society.

2. The contributions of women to household maintenance when the head was unable to provide food were more problematic. Though heads who were bearing their responsibilities fully sometimes actually paid cash to their wives for items that the wives contributed to the household food, such as groundnut oil, many wives apparently provided some of the

food regularly, without recompense. Thus in analyzing a family's nutritional status, for example, one may well have to take into account both men's and women's earning and purchasing capacity rather than carry the "men are responsible for food" generalities too far.

The ability to work. The fact that farm size in the Zaria villages increased directly with respect to the number of male adults available to work indicates that the physical ability to perform work is important.[15] The observation that male adults worked more hours in the labor bottleneck period is also indicative.

TABLE 5.5
Time Worked Per Family Male Adult, Zaria Villages, 1966-67

Variable Specification in Terms of a Male Adult	Dan Mahawayi	Doka	Hanwa		Average of the Three Villages
			Non-cattle Owners	Cattle Owners	
Time worked per year:					
Days: Family farm	140	159	125	118	140[a]
Off-farm	123	39	86	124	89[b]
Total	263	198	211	242	229[c]
Total hours	1287	971	1140	1378	1172[c]
Type of off-farm work (% of days):					
Traditional: Primary[d]	0.0	0.0	0.0	84.4	14.1
Manufacturing[e]	21.3	29.3	11.2	2.1	19.1
Services[f]	40.0	27.2	20.9	9.8	27.5
Trading[f]	35.0	24.7	3.4	0.1	20.5
Modern: Services[g]	3.7	18.8	64.5	3.6	18.8

[a] Average length of day worked on the family farm was 5.1 hours or 4.4 hours (excluding time walking to and from fields).
[b] The 20:80 percentile points for number of days worked by male adults were 161:317 days.
[c] The 20:80 percentile points for number of hours worked by male adults were 717:1489 hours.
[d] Involved looking after cattle.
[e] Included blacksmiths, tailors, carpenters, spinning, leather working and making pots, cigarettes, mats, sugar, etc. Average remuneration per day worked was N0.28.
[f] Included tending own house (fencing, building, thatching, cutting grass, and gathering firewood), barbers, butchers, hunting, begging, washermen, public officials, Koranic teachers, etc. Trading can also be classified as a traditional service. Average remuneration per day worked was N0.21.
[g] Included commission agents, messengers, laborers, night watchman, bicycle repairers, buying agents, etc. Average remuneration per day worked was N0.41.

Male adults in the survey villages each worked, on the average, almost 1,200 hours a year, including the time required to walk to and from the fields and including both farm and off-farm work. That labor was spread over about 230 days, indicating an average work-day of just over 5 hours (Table 5.5). The average time spent on the farm during the peak month, however, was 5.6 hours each day worked, including traveling time, compared with only 3.8 hours each day worked during the slackest farming month (Norman 1972). Until farm mechanization can substitute for some of that effort, the health and nutrition status of the working members of the household will continue to be a significant determinant of their ability to be productive. Though we did not measure individuals' health and nutrition status during the surveys, the consumption/expenditure information provided a notion of the adequacy of food intakes in relation to farm-work output.

Of the 109 households in both the farm-management and consumption surveys, 25 percent could be classified as having calorie intakes averaging below required levels, 12 percent as consuming just about the amounts needed, and 63 percent as having intake levels exceeding requirements by a wide margin. Several hypotheses to investigate the relationship between such intake levels and work effort suggest themselves. For example, household members whose food intake is lower than that required surely could be expected to be the least productive. Therefore, all other things being equal, members of such households quite likely would work fewer hours on both farm and off-farm activities, would emphasize occupations that are less effort-intensive, and would hire more labor on the family farm during peak periods than would members from households with higher per-capita food intakes.

One can also suggest that the analysis can usefully be turned in the other direction; that is, to look first at the work effort expended, then at the amount of output produced, and, finally, at the level of intakes that result. Such an approach, however, would require that one take into account an intervening decision-making procedure--that of choosing between retention of food for home consumption and that of selling it for cash. Based on that approach, then, food intakes would be not so much the determinants of the ability to work as the result of effort invested, subject to other decisions by the household head. The results of two regression models for examining some of the relationships between calorie intake and level and type of work, production, and income are given in Table 5.6. For the households in our survey, understanding the relationship between calorie intake per capita and the independent variables was complicated by a factor we discussed earlier: that a negative relationship existed between disposable income per capita and size of household, despite the fact that household size and total disposable income were positively correlated (Table 5.4). But incomes depend very much on the amount of work household members do.[16] So it should not be surprising that when expressed in total terms, incomes would show a negative relationship with per-capita calorie intake; hence, when expressed on a per-capita basis, individual members from large households surveyed had a lower level of welfare and therefore potentially lower per-capita calorie intakes than did members of smaller households. That conclusion is consistent with our earlier observation that per-capita calorie intake and household size were negatively related. However, when we expressed the independent variables in the models in per-capita terms, positive relationships, as anticipated, emerged between work, food production, or income per

TABLE 5.6
Relationship Between Calorie Intake Per Capita and Level and Composition of
Work, Production and Income, Zaria Villages[a]

(a) Specification

	Model	
Variables	A Agriculture	B All Occupations
Dependent	Calorie intake per capita	Calorie intake per capita
Independent:		
X_1	Food production[b] (kg per capita)	Disposable income per capita (N)
X_2	Family work on farm (man-hours per capita)	Total family hours work (man-hours)
X_3	Work with cattle (man-days per male adult)	
Dummy:		
V1	Hanwa = 1	Hanwa = 1
V2	Doka = 1	Doka = 1

(b) Results

	Model A		Model B	
Variable	B Value	Standard Error of B	B Value	Standard Error of B
Constant	1712.14		1871.33	
X_1	1.24[c]	0.87	13.85*	4.30
X_2	1.23[c]	0.97	- 0.07[d]	0.04
X_3	9.94	9.63		
V1	254.69	225.03	467.13*	219.21
V2	400.98*	197.89	462.21*	186.77
R	0.3175*		0.3746*	
Syx	798.69		777.17	

[a]Data used in this table included information from the farm management study
(1966-67) and consumption study (1970-71).
[b]Includes millet and sorghum production, the main food crops.
[c]Significantly different from zero only at the 25 percent level.
[d]Significantly different from zero only at the 10 percent level.
*Significantly different from zero at the 5 percent level.

capita, and calorie intake per capita.
    Nevertheless, the results of the models are not entirely
satisfactory.[17] The first model was designed to examine whether
level of food production and activity in agriculture, both crops and
cattle, were important in determining the level of food intake.
Although the signs were consistent with expectations, slightly
better results were obtained when per-capita calorie intake was
looked at in terms of all work and all income. This is perhaps not

surprising, because as we show later, most farming families engaged in a wide range of off-farm activities. Therefore, the welfare of households and individuals in them, and their resulting calorie intakes, would of course relate strongly to both sets of activities. The signs on the coefficients included in the second model were consistent with expectations.[18] A number of other models were also tested but added little to the results we present here.[19]

Resource use. As we have just discussed, family labor is one of the main resources possessed by farming households in northern Nigeria. But it is not the only one. Land, cash, cattle, and other capital assets are needed to complement household labor.

Earlier we differentiated two types of farmland: upland fields (gona), rainfed seasonally; and lowland fields (fadama), which can support crops throughout the year. Gona, by far the more dominant in the Zaria area, in 1966-67 accounted for 90 percent of the average farm of 3.9 hectares (Table 5.2). Mixed cropping and the ring cultivation system, practices discussed in Chapter 3, characterized the management of such land. On average, slightly less than 17 percent of the gona area was fallowed. Sole and double cropping systems were relatively more common on fadama. The limited quantities of fadama and the relative availability of water--due to the proximity of the water table to the surface--would logically lead to the conclusion that fadama would be used intensively. In fact, about 18 percent of the fadama was left fallow in 1966-67. A number of factors may have prevented land from being used more intensively. We give three examples:

1. Availability of labor to cultivate fadama may sometimes be important. For example, in 1966-67 a cultivated hectare of fadama required 137 percent more man-hours per hectare than did gona. In Hanwa, the need to look after cattle probably discouraged cattle owners from obtaining fadama (Table 5.2). In contrast, the apparent lack of off-farm opportunities in Doka was likely to have been one important factor encouraging the cultivation of fadama.
2. Flooding of fadama during the rainy season may discourage use of fadama throughout the year, as it did on some fadama in Dan Mahawayi and Doka in 1966-67.
3. Availability of market outlets for the produce from fadama is certainly an important factor. Many of the crops produced on such land were primarily cash crops of high value per hectare but of low value per unit weight and therefore expensive to transport. That no doubt contributed to the higher proportion of fadama left fallow in relatively isolated Dan Mahawayi, compared with Doka on the main Kano to Zaria road.

Turning to other resources, the average cost of capital used in crop production during the 1966-67 survey year was N20.60 per household, including hired (nonfamily) labor. Much of that amount, however, was based on imputed values of inputs and did not involve cash. Seed costs, for example, amounted to an average of ₦13.67 per year; up to 80 percent of the seed used was saved from the previous cropping year. Also 87 percent of the fertilizer cost was imputed; the total annual cost of fertilizer per household was about ₦5.67, but only ₦0.17 could be attributed to the use of inorganic fertilizer. Therefore, most of the fertilizer was in the form of organic manure, derived from livestock owned by the families or through contracts with nomadic Fulani cattle owners. Under those contracts, the manure produced on the field was often considered sufficient payment for the right of the Fulani to graze their cattle on the residues of the harvested crops.

Cash was used to obtain the services of inputs either on a temporary basis (e.g., renting, pledging, or leasing land, hiring labor, and purchasing seeds and fertilizers) or on a more permanent basis (e.g., purchasing equipment and the usufructuary rights to land). The average cash cost for crop production by families during the survey year--including cost for hiring labor--was only ₦25.15, which amounted to about 13 percent of the total value of production derived from crops in 1966-67. Cash expenses were, however, found to be very sensitive to overall income levels.

Less than 5 percent of the total cash expenses was, on the average, devoted to obtaining usufructuary rights to the land. About 13 percent was allocated to other durable capital investment, while only about 15 percent was for nondurable capital, consisting of seed and fertilizer.[20] The insignificance of marketing costs, which constituted only 3 percent of the total cash expenses, was related both to the relatively low proportion of total production sold--about 39 percent--and to the operation of middlemen or traders who often purchased products directly from the farming household and arranged for its transport to market.

Hiring labor, by far the most important item of cash expenditure on crop production, accounted for an average of almost 64 percent of the total cash expenses. The significance of this expenditure will become apparent in the following section, in which we show that labor was very limiting at certain times of the year.[21]

Two possible reasons why even more hired labor was not used in the Zaria survey year at peak periods were: first, as we briefly mentioned earlier, there was no class of landless laborers in the villages to fill that demand, so the time when hired labor was most in demand was also the time when everyone was busiest on their own farms; and second, more important from a resource perspective, few cash resources were available to farming families during peak periods which imposed a restriction on the amount of labor they could hire. Under such circumstances, particularly for those faced with cash-flow problems, there may not have been a great deal of potential for increasing the amount of hired labor.

Farm-production demands for labor. An important constraint to the productive employment of household labor throughout the year was the seasonal nature of rainfall, which--except when fadama land was available--largely restricted crop production to the rainy season. That implied substantial underemployment in the long dry season, especially since seasonal migration, which is important elsewhere in Nigeria, was not practiced in the Zaria area (Norman, Pryor, and Gibbs 1979). In 1966-67, about 39 percent of the total days worked by male adults in the Zaria villages were spent on off-farm work

(Table 5.5). What is perhaps unexpected about this off-farm work is that it was not concentrated only in the dry season. The potential for substituting between farm work and off-farm work, therefore, was perhaps not so great as would be desirable (Figure 5.3(c)). One, but by no means the only, reason may be that to be moderately successful in some off-farm operation during the dry season, a person had to provide some continuity to that commitment during the year. That was particularly true for occupations that involved regular clientele. For example, villagers in the Zaria area who were cattle owners and those engaged in crafts and services such as trading obviously had year-round commitments.

The degree of seasonality in crop production is illustrated by the values of the coefficients of variation calculated with respect to the number of man-hours spent per month on the family farm. In the Zaria area, depending on the village, they ranged from 42 to 55 percent during 1966-67 (Norman, Pryor, and Gibbs 1979). Agricultural activity usually peaked in June (Figures 5.2 and 5.3). An average of 256 man-hours per month was spent on the family farm during this peak month, 70 percent more than the average monthly input of 150 man-hours. March was usually the slackest month. The labor input on the family farm during March amounted to only 34 man-hours, 77 percent fewer man-hours than the average monthly input. The disparity in the monthly distribution of labor on the family farm was emphasized even further by the fact that the four busiest months of the farming year (May to August) accounted for more than 50 percent of the total annual labor input, whereas the four slackest months (January through April) accounted for only 16 percent.

The seasonality of crop production led to the conclusion that the amount of land that a family could work during the labor bottleneck period fundamentally determined the level of agricultural activity during the rest of the year. At the time of the study, it should be kept in mind, the power base was hand labor and virtually no improved technology had been adopted, so the major labor constraint occurred during crop cultivation which included thinning, weeding, and ridging activities (Figure 5.2). Virtually all the farming households used three methods in attempts to ameliorate the worst effects of the labor bottleneck period: adjusting farming practices, increasing family labor inputs, and increasing the use of hired labor.

One obvious example of a farming technique used by nearly all farmers was early weeding, before the bottleneck period. That permitted further weeding on those fields to be postponed until well into the period. Much of that weeding often was done before it was really necessary, but it was a rational response to the anticipated rise in the opportunity cost of labor as the bottleneck period approached. As we showed earlier (Chapter 3), another common way of alleviating the adverse effects of a labor shortage was to grow crops in mixtures.

Even then, family labor inputs often had to be increased; reallocation was insufficient on its own. As stressed earlier, women's seclusion precluded female adults from contributing much to work on the family farm. In addition, the labor inputs of older children represented only a small proportion of the total. Now that

120

FIGURE 5.2
Monthly Profile of Activities on the Family Farm,
Zaria Area, 1969-74[a]

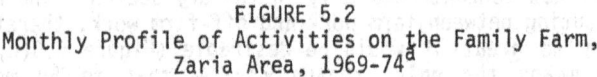
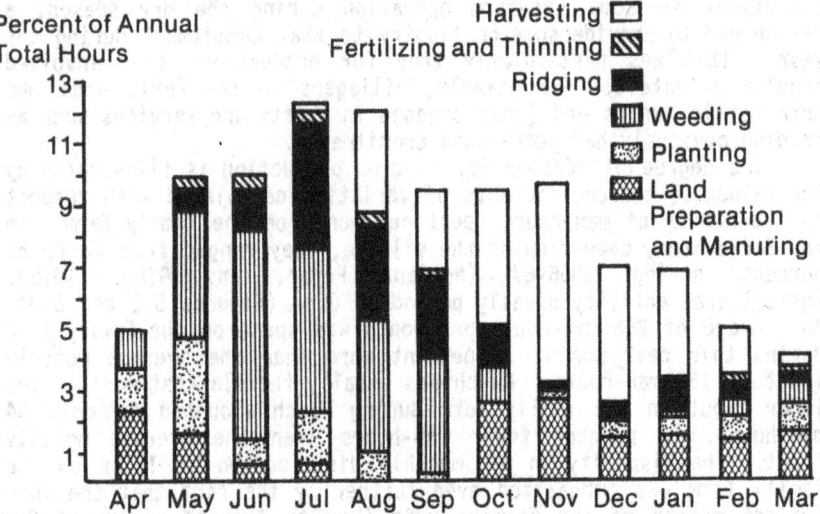

[a]This is an average for seven farming families over a five year
period (Roth 1979). Lower than average annual rainfalls in the
early 1970s probably partially accounted for a later than average
bottleneck period in July and August.

education has become more widespread--through the Universal Primary
Education program--the labor input of older children likely will
become even smaller than it has been. Therefore, it is apparent
that family male adults will continue to provide the major input on
the farm. As observed in our study, during June, the peak
production month, a male adult worked about twenty-four days at all
jobs, as opposed to nineteen during an average month; that meant
that he spent about 26 percent more days working in June than in the
average month. But, even when farm-labor demands were at the peak,
a male adult spent only an average of seventeen days working on the
family farm, allocating seven days to off-farm work (Figure 5.3).
In addition to the continuity reason cited earlier, one must also
note the importance of off-farm work as a source of cash for many
farmers. In savanna agriculture, little income is obtained from
farming activities until after the bottleneck period is over. Cash
and food resources tend to be low because most crops are harvested
between August and December. Therefore, the farming
households--usually those with small farms--facing severe depletion
of cash and food resources are compelled to work in off-farm
employment even though the work needs of their own farms might be
high (Matlon 1977).

This cash shortage helped to explain as well the differing
patterns for using hired labor. One would expect that since labor

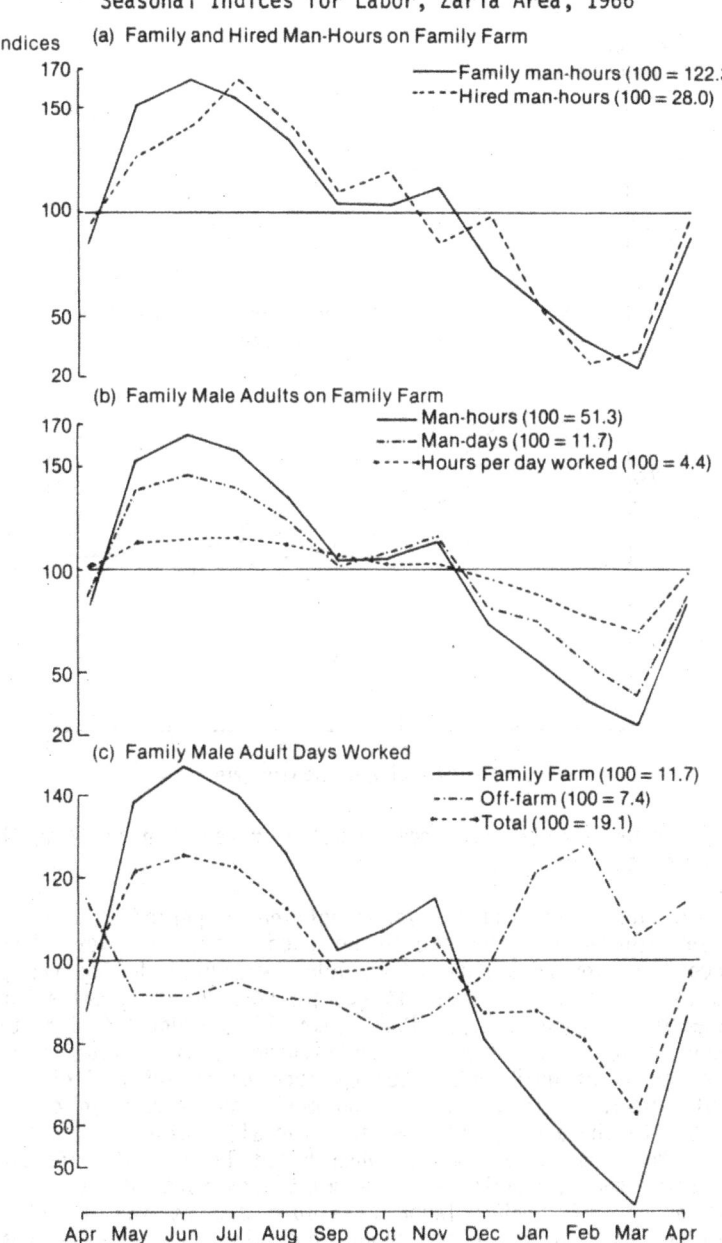

FIGURE 5.3
Seasonal Indices for Labor, Zaria Area, 1966[a]

[a]The indices represent the average of the whole sample in the three survey villages.

122

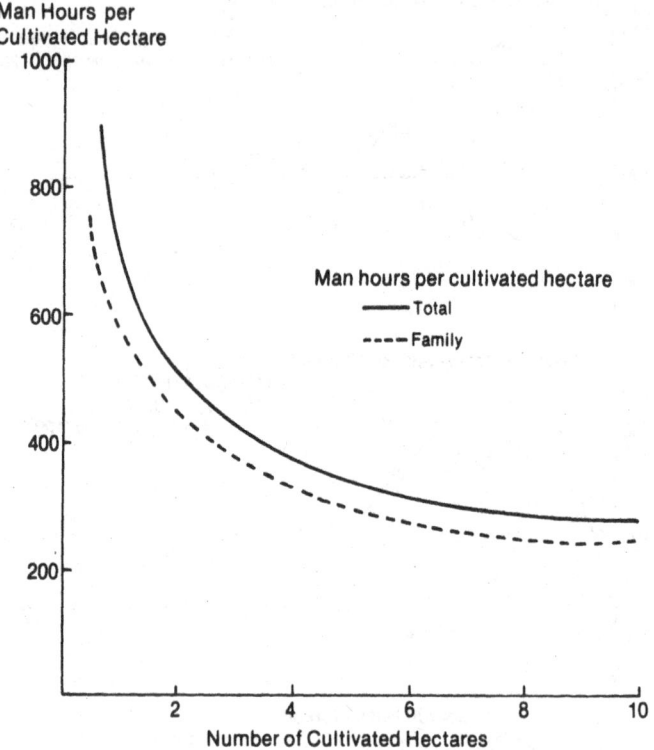

FIGURE 5.4
Relationship Between Labor Input per Cultivated Hectare
and Number of Cultivated Hectares, Dan Mahawayi, 1966-67[a]

[a]This graph was constructed from functions given in note 22 at the
end of the chapter.

is in such demand during the labor bottleneck period, the bulk of
hired or non-family labor would be used then. However, despite
evidence that somewhat more hired labor was used during the peak
period in the study villages, it was perhaps not so great as would
be expected (Figure 5.3c). More than 18 percent of the total
man-hour input on the farm was contributed by hired labor and the
greater amount of work undertaken by such hired labor during June
and July involved longer hours and more days--seven or eight in
contrast with the average of five days for all months.

But to understand how and when hired labor most effectively
supplemented the household labor, we must look more deeply not only
at the amount of family labor available and at the household's
ability to pay cash for hired labor, but also at the relationship
between labor and land. It is reasonable to suppose that the total
amount of labor used per cultivated hectare would be inversely
related to the total number of cultivated hectares on the farm,
given equivalent quality of land. Indicators of such quality as the

proportion of cultivated land that was <u>fadama</u> and the amount of organic manure applied per hectare imply that not all land might have been of equivalent quality. It is likely, therefore, that the higher the quality of land, the greater would be the number of man-hours devoted to it on a per hectare basis. Graphs drawn from regression models verifying such relationships are shown in Figure 5.4. But the models indicated that the level of family input per cultivated hectare decreased less rapidly than did total man-hours per cultivated hectare as the number of cultivated hectares on the farm increased. The difference could, of course, be attributed to the use of hired labor. The total number of man-hours of hired labor used by a household was in fact shown to be significant and positively related to the number of cultivated hectares.[23] The significant relationship, however, did not hold when hired hours were expressed per cultivated hectare. That implies that the use of hired labor did not offset the decrease in family labor inputs per cultivated hectare as the number of cultivated hectares increased.

Farming households are obviously interested in relating labor inputs to production. In the context of the current discussion on relating labor to land, the results estimated for Dan Mahawayi in Table 5.7 indicated that the marginal productivity of family labor was greater on the larger farms, due, as expected, to greater areas available per unit of labor and therefore lower levels of labor input per unit of land. The much higher hired-labor inputs used on larger farms, probably due in part to more acute seasonal cash shortages by families operating small farms--which precluded their hiring more labor--could have contributed to the higher marginal productivity of hired labor estimated for small farms.

<u>Opportunities for off-farm work</u>. Opportunities for off-farm employment in the survey villages were found to be related to the location and accessibility of the village. Employment in the traditional sector consisted of those jobs that were fairly independent of the developmental process; in other words, they were jobs that had been undertaken for many generations. In contrast, jobs in the modern sector were those arising directly or indirectly as a result of improved communications and the development of large cities, commercial firms, and government organizations. In the case of Hanwa, the village most accessible to Zaria, families living there not owning cattle generally found jobs in the modern sector, which usually were more remunerative than work in the traditional sector (Table 5.5). Although the relative isolation of Dan Mahawayi precluded residents there from obtaining modern jobs associated with Zaria, a substantial number of traditional services still flourished. In Hanwa, persons interested in part-time traditional occupations had to compete with full-time specialists and those employed in industries producing modern-substitute products in Zaria. In Doka, the village with intermediate access, employment opportunities in the modern sector were quite limited; at the same time, traditional activities also were reduced because of accessibility to, and thus competition from, the urban sector of Zaria. On the positive side, however, Doka's accessibility to a main road encouraged farmers to cultivate the remunerative sugarcane on <u>fadama</u>, thereby increasing their incomes and decreasing their

reliance on the often less certain sources of off-farm activities.

TABLE 5.7
Marginal Value Products of Land and Labor on Small and Large Farms, Dan Mahawayi, 1966-67[a]

Variable Specification	Small farms		Large farms	
	Average Level of Input	Marginal Value Productivity	Average Level of Input	Marginal Value Productivity
Land (cultivated hectares)[b]:				
Gona	1.81(480)	21.09	5.00(330)	18.55
Fadama	0.14(839)	14.11	0.32(1300)	15.01
Labor (man-hours):				
Family	909	0.03	1223	0.06
Hired	128	0.06	689	0.03

[a]Small farms were defined as families having farms with land per resident ratios of less than 0.6 ha, whereas large farms were those with land per resident ratios more than 0.6 ha. The marginal value products were estimated at the average input levels for small and large farms in Dan Mahawayi by using the production function given in Table 5.12.
[b]Figures in parentheses indicate the man-hours per cultivated hectare.

Incentives to work. Food needs are perhaps the greatest incentive for rural Zaria farming households to work. It is imperative that food be provided on a daily basis and in adequate amounts and that lies behind a whole range of household decisions. The single growing season associated with the savanna climate encourages a certain amount of planning ahead to provide staples and grain for the whole year. The predominance of sorghum and millet in production patterns was closely linked with their roles as staples in household diets. On the average, about 90 percent of the volume of these grains reportedly consumed in the 120 sample households in the 1970-71 survey year was from own-farm production. Security of the household food supplies was exhibited not only in terms of decisions as to what crops to produce but also in terms of the practices used in their production. For example, the practice of mixed cropping was found to be, as we showed earlier, consistent with the notion of security. Fortunately, under indigenous technology conditions, that mode of production also resulted in greater total output per unit of land and labor. Thus, the security objective of Zaria farmers did not imply, under their current cultivation systems, a negative trade-off with total production. Another production determinant was the supply of household labor; a larger labor force permitted the farming household to cultivate a

TABLE 5.8
Food Consumption, Own-Farm Production and Relative Costs, Zaria
Area, 1970-71[a]

Item	Average Daily per Capita Consumption		Kobo per[b]	
	Calories	Value	1000 Calories	100 grams Protein
Cereals:	1587(88)	3.4	2.1	9.6
Sorghum	1178(90)	2.3	2.4	8.8
Millet	274(98)	0.5	2.2	9.3
Maize	78(81)	0.1	2.7	11.2
Cereal products	218(14)	0.6		
Starchy roots	48(44)	0.2	4.8	55.2
Seeds, nuts, legumes:	89(21)	0.5	5.8	9.2
Cowpeas, raw	24(61)	0.1	3.2	5.3
Groundnuts	7(44)	0.1	2.4	5.3
Oils and fats:	194(2)	0.4	2.5	
Groundnut oil	32(2)	0.1	2.9	
Palm oil	160(0)	0.3	2.3	
Meat	19(2)	0.8	42.0	31.3
Fish and poultry	3(19)	0.1		
Milk	23(25)	0.2	12.0	16.6
Vegetables: fresh	14(48)	0.4	28.0	91.7
dried	26(41)	0.2	9.7	25.3
Fruits	3(42)	0.1		
Sugar, sweets	9(15)	0.1	6.4	
Salts, spices	1(0)	0.1		
Snacks, miscellaneous	19(3)	0.4		
	2253(67)	7.5(47)	3.3	12.2

[a]Figures in the table represent the average of the three villages;
those in parentheses the percentage produced on the family farm.
[b]When information was not available, gaps were left in the table.

greater area and thus to increase total output. The greater labor
force, however, placed greater demands on the household's
farm-produced food supply, especially if each active worker was
accompanied by nonfarming dependents--women and small children. In
that case, the production objective and the goal of food security
seemed to be at odds. In this and later sections we look at how
crop-production processes, sales, and food-purchasing behavior were
related to the food needs implied by the consumption patterns of the
households in the Zaria villages.

Table 5.8 shows a comparative perspective on average food
consumption in the three villages by contrasting daily own-farm
consumption with total consumption on a per-capita basis. The
consumption patterns in all three villages indicated substantially
the same types of diet, although some differences were found for

individual items. Determinants of the consumption patterns of such items were partly a function of the level of production on the family farm and the reliance on the market place for providing the item. For example, Hanwa's lack of <u>fadama</u> land but its accessibility to the urban markets, as well as the dynamics of household food purchasing, probably combined to account for the low levels of home-produced vegetable and fruit consumption in Hanwa compared with the other villages. The figures in Table 5.8 also indicate that in the survey villages there was on average substantial but not complete self-sufficiency in food production. While 67 percent of the average calorie intake was derived from own-farm production, the foodstuffs comprising that amount account for only 47 percent of the average monetary value of food consumed daily.[24] That so much cash can be exchanged even in very rural areas in northern Nigeria has been noted by a number of authors (Hill 1972). That a money economy rather than a strictly subsistence orientation characterized the three Zaria villages should, therefore, be no surprise. Much of the food eaten for the first two meals each day was purchased in a ready-to-eat form. Many of the ingredients of the evening meal were also purchased, particularly those for the soup.

In the 1970-71 consumption study, we found the value of food produced on the family farm to be 2.2 kobo (₦0.022) per thousand calories and purchased food to be 5.5 kobo per thousand calories. Table 5.8 provides a basis for comparing the relative costs of various foods in nutritional terms, although critical information on several purchased ready-to-eat items was not available. The rationale for purchasing more grain supplies to supplement production shortfalls is apparent: cereals ranked among the lowest-cost commodities in all villages for calorie supplies, and only legumes were more economically efficient suppliers of protein. Milk in both Hanwa and Dan Mahawayi was a relatively cheap source of protein, but its high cost in Doka probably helped to account for its lower consumption there. The importance of palm oil and groundnut oil as sources of calories in the diet was confirmed by the cost figures for those commodities, which also partially explained the preference for palm oil.[25] Though it was somewhat unexpected that the cost of locally produced groundnut oil would be slightly higher than that of palm oil--from southern Nigeria--the relative costs of other food commodities in general reflected transport, production, and perishability factors.

As 67 percent of the calories were obtained at no cash cost, the overall average value of food consumed by an average person each day was 3.3 kobo per thousand calories. Thus, the average household spent approximately ₦123 cash on food in 1970-71, representing more than 30 percent of all household cash expenditures during that year.

These averages, of course, gloss over the substantial differences among villages and households. Given the average-sized household in each village, for example, annual estimated cash expenditures for food alone were about ₦80 in Doka and Dan Mahawayi, but approximately ₦220 in Hanwa. One household in Dan Mahawayi, however, purchased more than 99 percent of the value of food it consumed, whereas some in Hanwa paid cash for as little as 27 or 28 percent. Average satisfaction of food requirements--estimated in terms of calories and protein intakes needed for the age and sex composition of each household--was greater in Doka and Hanwa than in Dan Mahawayi. In each village, however, a number of households reported calorie intake levels below estimated needs. In examining the relationship of such nutrient intake levels to a household's expenditure of physical work effort, we also suggested that work time expended might have had a positive impact on the household's ability to meet its food requirements. That latter relationship was complicated by several intervening decision variables relating to other cash needs, commodity preferences, and other factors more difficult to measure.

The classic example of farming families supposedly selling crops at harvest to pay off production debts and buying crops at higher prices later in the year to feed their families illustrates one such variable. However, while not denying that may occur in many parts of the savanna, we shall show later in this chapter that did not occur in the Zaria villages. Thus, the sales and purchasing behavior of the farming families in the Zaria villages helped explain the economic rationality of the village decision-makers. Information on returns to various possible activities demonstrated that Zaria farmers tended to allocate their work time to those activities having positive economic incentives. For example, in our earlier comparative analysis of crop mixtures and sole crops, we concluded that from an economically rational viewpoint farmers were justified in devoting their limited land or labor to crop mixtures, which continue to dominate in the area.

Perhaps as a final point, however, we should note that one should be careful not to overstate the complete explanatory power of either food or cash needs as the motivating force behind rural work-time allocation. A linear programming model used to explore the possibility of greater profit-maximizing behavior among Zaria farmers indeed indicated that an extra 280 man-hours expended in producing more millet and sorghum could have increased the net income from crops of a typical household by almost 18 percent (Norman 1970). Unfortunately, it was not possible, given the data-base available, to maximize other variables, such as the need to invest in the bride price for a new wife, and these remain hypothetical rationales to explain observed behavior.

Household Wealth, Investments, and Other Assets

Inheritance, investments, and wealth. The role of inheritance in providing farmland has already been discussed. A young man normally had to inherit a certain amount of land as well as other capital if he intended to start farming on his own, especially if he had been working in a land-poor gandu. If off-farm jobs were available, they could provide part or all of the capital needed, however. In Hanwa, for example, farmers were able to acquire the means to farm without inheriting them.

Durable investment needed for crop production in the Zaria area was low. Dependency on hand tools, together with the absence of farm buildings other than grain stores and an occasional livestock hut, resulted in an average inventory value of investment of only N4.51 in 1966-67. The close linkage between farm operations and household existence, however, meant that a young man desiring an independent farming operation needed not only a farm capital, and land, but also a wife--which required bride-price payments--and living quarters. Again, normally these involved inheritance of land and capital, with a father providing some of the payment for a son's marriage before the son left the father's farming operation, and the inheritance of living quarters. With off-farm jobs, however, a young man might earn enough on his own to afford the prerequisites for farming, but it was difficult.

Setting up a small-livestock enterprise also may require substantial investment, although inheritance was not usually the way this investment was acquired in the Zaria area. Ownership of small animals, including poultry, normally was acquired gradually, on an individual basis with both investments and returns handled personally by the owners. Whereas only men could easily acquire land, livestock ownership was open to women. Ownership of cattle, however, was tied to ethnic background (Fulani), to inheritance traditions, and possibly to the existence of a household unit large enough to manage the herding, although other herding arrangements (riko) might sometimes be made.

But it should be noted that investing in livestock of all types --chickens, sheep, goats, guinea fowl, donkeys, and horses--was relatively significant in the village households, despite the fact that such livestock did not play an important role in farming activities, food supplies, or annual incomes of most households. The role of livestock in wealth accumulation and inflation-proof savings no doubt contributed to the popularity of owning livestock as an investment. Livestock could be readily translated into cash when needed, but until then could bear interest--in the form of products and offspring, as well as manure. In 1966-67 the few households owning cattle had an average investment in cattle of ₦604.08; the average level which all households invested in other livestock was ₦15.62.

Acquiring skills and productivity. Although education and literacy are somewhat intangible assets, compared with land, in the Zaria context the quality of skills possessed by household decision-makers is in fact a crucial asset.

Hardly any household heads were literate in Hausa, although some could read the Koran in Arabic. Their means for acquiring a new skill were visual (i.e., watching somebody else), aural (i.e., hearing about it), or experimental (i.e., simply trying something out). Thus, the acquisition of skills was to a great extent a function of experience and, to some extent, age. None of the villages were regularly serviced by extension agents from the Ministry of Agriculture and Natural Resources and few farmers had ever seen or talked with one; new inputs of farming skills were thus minimal. However, as the years of our research involvement progressed, several new inputs were introduced in the Zaria villages, sometimes with instructions delivered by the enumerators. Some of those innovations proved worthwhile, with opportunities for farmers in the villages--other than those to whom the innovation was introduced--to observe and ask about the innovations.

Women confined to the compounds obviously had even more

impediments in the way of acquiring skills. Indeed, most of the few new skills that we observed had been learned by women when they were in other villages or towns visiting relatives. Some women, however, were also instructed in new skills by their husbands; in Hanwa, for example, the use of powdered baby milk was said to be the result of a husband's encouragement and wishes.

In this rather limited learning environment, then, it was interesting to note the differentials among people in seizing the opportunities that did exist. One woman in Hanwa, for example, was informed by her husband that a major new construction was beginning in Zaria. She took his advice of preparing lunch food for sale at the construction site, employing her son as a retailer. Even though her workload was significantly increased, she believed the extra effort was worthwhile because her cash flow was improved.

Endogenous Influences: a Summary

In the preceding section, we have outlined some of the endogenous influences that constitute the decision-making variables in rural Zaria households. The successful farming household was one that was able to do the following: marshall enough labor to cultivate its land, but not to the extent that there would be too many mouths to feed from too little land; earn enough cash to provide additional food and labor to supplement household resources; and acquire the assets that would permit growth and diversification of the household enterprises. For these reasons, farming systems researchers need to look not only at land per household, but at land per laborer; not only at production per household, but at production per capita, or perhaps amount available for consumption per capita; not only at cash incomes per se, but at cash earnings in relation to expenditures.

We now turn to the production processes that visibly reflect the choices made by rural decision makers: crops, livestock, and off-farm work.

THE CROP PROCESS

Crops Grown

In the study villages, cereal crop production accounted for 51 percent of the total adjusted hectares planted during the 1966-67 rainfed growing season and contributed 46 percent of the total value of the crop production (Table 5.9). Analogous figures for grain and legumes were 21 and 18 percent, respectively, with the remaining contributors being starchy roots and tubers, vegetables, sugarcane, and nonfood crops. The principal crops grown on the rainfed upland were sorghum (guineacorn), millet, cowpeas, sweet potatoes, groundnuts, and cotton--the latter two largely in the status of cash crops. Sugarcane, grown solely on the fadama, was also a cash crop. Of the twenty-five crops grown widely in the study villages, only five or six could be termed major crops in an economic sense. Minor crops, such as cassava, okra, pepper, maize, rice, and onions played important cultural or social roles or were critical ingredients in the diet, but were in the aggregate a small part of total commercial

TABLE 5.9
Adjusted Hectares and Value of Production By Crop Class, Zaria Villages, 1966-67[a]

Crop Class	Percent of Adjusted Hectares				Percent of Crop Value of Production			
	Dan Mahawayi	Doka	Hanwa	Average	Dan Mahawayi	Doka	Hanwa	Average
Cereals	51.6	47.1	55.7	51.5	40.7	36.4	61.6	46.2
Grain legumes	23.8	22.1	18.6	21.5	24.7	18.1	12.7	18.5
Starchy roots and tubers	6.1	5.3	10.3	7.2	5.2	1.4	9.8	5.3
Vegetables	3.0	7.2	3.1	4.4	5.7	6.2	3.0	5.0
Sugarcane	3.7	12.0	1.9	5.9	16.4	35.2	6.9	19.5
Nonfood	11.8	6.3	10.4	9.5	7.3	2.7	6.0	5.5
Total:								
Adjusted hectares	3.8	2.9	2.8	3.2				
Value of crops (N)					197.2	187.9	212.2	199.1

[a]The calculation of adjusted hectarage was necessary because of extensive use of mixed crops. The adjusted hectarage of each crop in the mixture was calculated by dividing the hectares devoted to the crop mixture by the number of crops in the mixture. For example, a two hectare millet/sorghum mixture was recorded as one hectare of millet and one hectare of sorghum.

transactions. Each household grew between four and fifteen crops, with eight crops being the average. The relative emphasis on various crops in Hanwa and Dan Mahawayi was in general very similar (Table 5.9). However, because of the fadama and accessibility, farming families in Doka tended to forego production of some cereal crops in favor of sugarcane. Because sugarcane was highly productive on fadama, in per-hectare terms in Doka, it contributed 35 percent of the total value of crop production obtained by an average farming family while accounting for only 12 percent of the area cultivated by the family. Besides this obvious example, what other factors influence crop choice and productivity? These are key issues of particular importance to FSR workers who are seeking ways to augment the welfare of farming families through increasing the productivity of farming systems in ways that are acceptable to the households. We now briefly look at these two issues in terms of the farming households in the study villages.

Crop Choice

Invariably, adopting improved technology involves greater commercialization of agriculture. At the very least, it usually involves disposing of some of the increased production in the market place, and, more often than not, the adoption of improved technology also necessitates purchasing some of the inputs. What route does this commercialization take? Obviously, the answer is very location specific. We now briefly examine, in the context of the Zaria villages, three hypotheses concerning possible factors influencing whether or not farming households would sell crops, which in turn would reflect the types of crop produced. However, to put this discussion in the proper perspective, it is important to remember that in the Zaria area it is technically possible to grow both food crops (such as millet and sorghum) and those that are usually termed cash crops (such as groundnuts, cotton, or sugarcane), and also that during the time of the farm-management survey, 1966-67, very little in the way of improved technology had been adopted in the survey villages.[26]

Food needs. As we have shown earlier, cereal crops (that is, millet and sorghum) appeared to dominate in both the production and dietary systems of farming families. Therefore, the first hypothesis involved testing the relationship between the amount of food produced and family or household size. As shown in Table 5.10, the expected significant positive relationship between the two variables did exist when we took account of village location.[27] It is interesting that the signs on the location variables verified the lower level of food production in Doka, compared with that in Hanwa or Dan Mahawayi.

Because of our earlier finding that the per-capita calorie intake was negatively related to household size, we examined the above relationship between food production and household size to see if it would be maintained when food production was expressed in per-capita terms.[28] The relationship was no longer significant; therefore, it provided some support for our earlier conclusion that

TABLE 5.10
Factors Influencing Food Production and Sales of Crop Production, Zaria Area, 1966-67[a]

Independent Variables		Dependent Variable			
		Food Production (kg)[a]		Value of Crop Production Sold (₦)	
		b Value	Standard Error of b	b Value	Standard Error of b
Constant		304.3026		54.4518	
Size of household	X_1	148.8283*	19.6439	0.4672*	0.1096
Family work on farm (man-hours)	X_1				
Dummy variables[b]:					
	X_2	155.3594	251.7453	-810.6897*	253.8393
	X_3	-330.8959	230.9312	-284.5774	238.5507
R		0.6188*		0.3997*	
Syx		1064.9912		1081.4984	

[a]Calculated as the sum of millet and sorghum production.
[b]These are defined in Table 5.6.
*Significantly different from zero at the 5 percent level.

household welfare was not necessarily synonymous with individual welfare.

Diversity and cash-crop emphasis. Although the need for food appeared to be very influential in determining what crops were grown, it is possible that the decision as to whether or not to produce food crops was complicated by other considerations. Two additional hypotheses are as follows:

1. With greater diversity of occupations--such as looking after cattle and doing off-farm work of various types in addition to farm work on crop production--there is less risk in entering the market place, and therefore more of the crop produced likely will be sold.
2. The greater the emphasis that is placed on cash-crop production--that is, the less the emphasis that is placed on major food crops--the greater will be the value of crops sold.

The first hypothesis was not verified by the results (Table 5.10).[29] As might be expected, the value of sales of crops was closely related to time spent working on the farm. Other variables reflecting time devoted to off-farm work and looking after cattle, however, were not significant and therefore were excluded from the model. In any case, even this farm work variable lost its significance when expressed in per-capita terms, and when in addition the value of actual sales of crops was also computed in per-capita terms.

Therefore, diversity of occupations, at least at the level of farming families in the Zaria villages, did not appear to encourage commercialization or selling of crops. Perhaps one factor influencing that is that farm or crop production work still consumed a major part of the male adult's time in the study villages (Table 5.5).

Turning next to the mix of crop enterprises, what about the influence of placing greater emphasis on cash crops, including minor crops? The results of the models given in Table 5.11 include a variable denoting cultivated area because the potential for a larger variety of crops increased with an increase in area cultivated. The significant negative sign on the diversity-index variable confirmed the notion that relatively greater emphasis on crops other than the major food crops would result in greater sales of crops. As would be expected, a significant positive relationship existed between sales of crops and cultivated area, and interestingly, that degree of significance was maintained when the variables for sales and cultivated area were expressed in per-capita terms.[30]

In both models, the locational variables indicated that the value of crop production sold in Doka, all other things being equal, was greater than that in either of the two other villages, presumably because of the greater emphasis on sugarcane in Doka.

Cultivated area. As noted in the preceding model, the cultivated area was important in determining the value of crop production sold. But would an increase in cultivated area indicate

TABLE 5.11
Relationship Between Level of Crop Sales and Relative Emphasis on Cash and Minor Crops, Zaria Area, 1966-67

Independent Variables		Dependent Variable			
		Value of Crop Production Sold (N)		Value of Crop Production Sold/Resident (N)	
		b Value	Standard Error of b	b Value	Standard Error of b
Constant		45.0681		10.4917	
Cultivated hectares:					
Total	X_1	34.0757*	1.9851	22.2564*	2.9744
Per resident	X_1			-58.1434*	28.1475
Crop diversity index[a]	X_2	-419.7469*	181.6603		
Dummy variables:					
V1	X_3	- 20.0791	14.0304	- 1.6828	2.3369
V2	X_4	15.4132	13.8428	3.5475	2.1986
R		0.8534*		0.6368*	
Syx		61.7538		9.5715	

[a] The index was calculated as follows:

$$\sum_{i=1}^{n} \frac{(A_{im} \cdot P_i)}{C_m}$$

where: A_i = Adjusted hectares of crop i grown by household m.
P_i = Proportion of adjusted hectares devoted to crop i on an average farm in the Zaria villages.
C_m = Hectares cultivated by household m.

[b] The dummy variables are defined in Table 5.6.
*Significantly different from zero at the 5 percent level.

FIGURE 5.5
Relationship Between the Proportion of the Value of Crop
Production Sold and Area Cultivated, Zaria Villages, 1966-67[a]

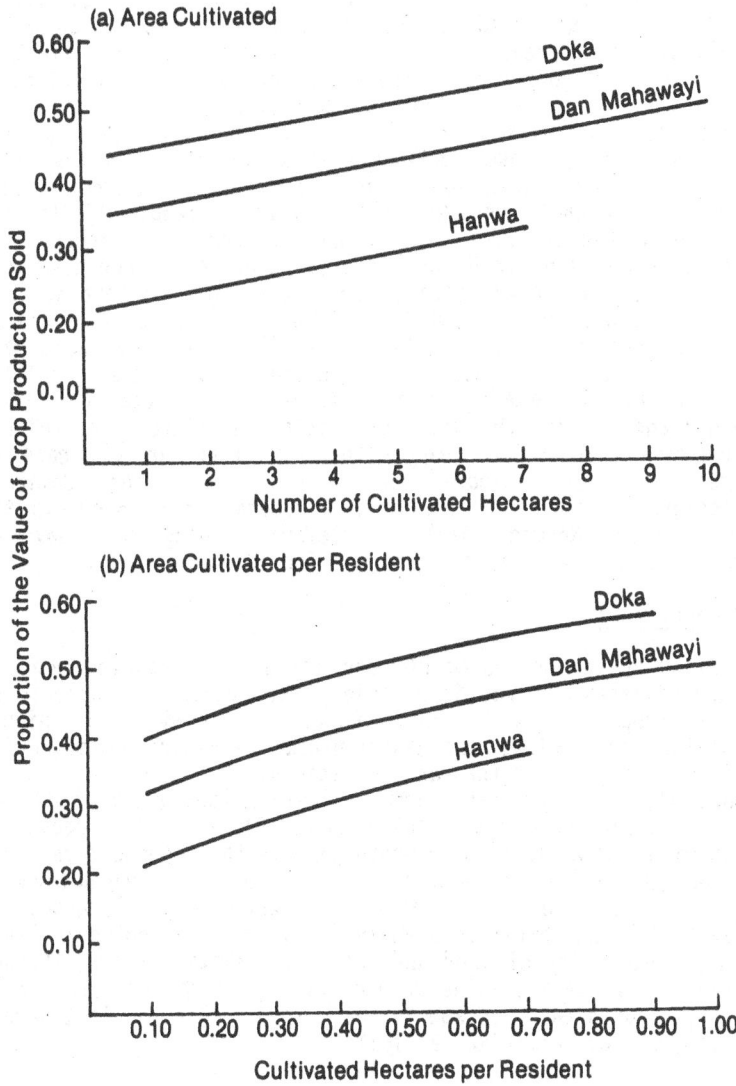

[a]The graphs were estimated from functions given in note 31 at the
end of the chapter.

a concomitant increase in the farmer's relative emphasis on
producing crops for sale? The results of the best-fit models that
we estimated and which are graphed in Figure 5.5 indeed verified the
hypothesis that the degree of market orientation--as measured in
monetary value, by the proportion of produced crops sold--increased

in the study villages as the cultivated area increased. Moreover, that relationship was maintained when the area cultivated was expressed in per-capita terms.

That conflicted with results obtained in neighboring areas by Matlon (1977) and Balcet and Candler (1981), who suggested a U-shaped relationship between proportion of crop-production value sold and cultivated area. In other words, families with little land marketed a higher proportion of their crop production than did those with medium-sized farms, who in turn marketed proportionately less than did those with large farms. One reason suggested for the unusual behavior of those families with small farms was that they needed to sell off production to pay back loans. Another reason might have been that, because of the limited land, the farming families grew higher-value cash crops--in terms of return per hectare--that could be sold, and used the proceeds to purchase more food than could have been obtained from devoting that land to food rather than cash crops (Matlon 1977). We can offer no satisfactory explanation for the difference in our results except to emphasize, as we do later in the chapter, that, in the Zaria villages, selling food crops did not seem to be tied to production credit. Also, groundnuts and cotton, the major cash crops on rainfed gona, either were usually grown in mixtures with food crops--in the case of groundnuts -- or compared unfavorably with the dominant millet/sorghum mixture in terms of net return per hectare--in the case of cotton (Norman 1972). Therefore growing them gave no comparative advantage over growing just food crops.

Crop Productivity

Because of the dominance of crop mixtures on rainfed gona in the study villages, it was impossible to do meaningful analysis on individual crops. We did show in Chapter 3, however, the higher productivity per hectare and per man-hour resulting from growing crops in mixtures rather than as sole stands.

What then, in general terms, influenced the productivity of crops? In Table 5.12 the results of a Cobb-Douglas production function with the value of crop production as the dependent variable are presented together with the estimates of the marginal value productivities of the various inputs estimated at their geometric mean levels for the sample as a whole.[32] We have presented in Table 5.13 the productivity of land and labor, in relative terms, through estimating their average-value productivities at the geometric mean levels existing in each village for each input. The results, hardly surprising, can be summarized as follows:

1. Overall, the average productivity of fadama was much greater than that of gona, reflecting the greater availability of water, thereby permitting the cultivation of sugarcane.
2. Overall, the average productivity of hired labor was much greater than that of family labor, reflecting both its more limited use and its use only when absolutely necessary, because it usually involved explicit payment of wages.
3. With reference to land, gona was most productive in Hanwa, partly because of the highest level of labor input per

TABLE 5.12
Production Function for Value of Production Derived from Crops,
Zaria Area, 1966-67[a]

Independent Variables		Coefficient		Geometric Mean[b]	
		Value	Standard Error	Value	MVP
Constant		0.8363			
Cultivated hectares:					
Gona	Log X_1	0.3360	0.0665	2.15	23.35
Fadama	Log X_2	0.0174	0.0378	0.45	5.75
Man-hours of work by:[c]					
Family	Log X_3	0.2728	0.0684	1195.09	0.03
Hired labor	Log X_4	0.0639	0.0170	46.57	0.20
Other inputs:					
Fixed costs	Log X_5	0.0345	0.0442	1.62	3.18
Variable costs[d]	Log X_6	0.2400	0.0424	8.85	4.05
Dummy variables[e]:					
V1	X_7	0.0730	0.0394		
V2	X_8	0.0832	0.0374		
R		0.9050			
Syx		0.1458			

[a] A Cobb-Douglas function was estimated with value of crop production expressed in naira (₦).
[b] These of course deviated significantly from the arithmetic means used mostly in the chapter, and reflected the geometric mean for the whole sample.
[c] Excluded time travelling to and from fields.
[d] The variable costs excluded funds used for hiring labor which were in essence accounted for in variable X_4.
[e] For the definition of the dummy variables see Table 5.6.

cultivated gona hectare. Fadama, heavily cultivated in sugarcane, was most productive in Doka.
4. The spreading of the lower family labor level over a larger cultivated area no doubt helped bring about a higher productivity of family labor per man-hour in Dan Mahawayi compared with that in Hanwa. However, hired labor was most productive in Doka where lower levels were employed than in either of the other two villages.[33]

Further insights into the factors determining the productivity of land and labor are possible through examination of the results in Table 5.14. Not surprisingly, the proportion of the cultivated land that was fadama, together with the level of labor input, was significant in determining the gross return per cultivated hectare.

TABLE 5.13
Average Value Productivity of Land and Labor at Geometric Levels of Input Use, Zaria Villages, 1966-67[a]

Input	Geometric Mean Level of Input Use (Units of Input)				Average Value Productivity of the Input at the Geometric Mean Level (₦ per Unit of Input)			
	Whole Sample	Level as Index of Whole Sample			Whole Sample	Level as Index of Whole Sample		
		Dan Mahawayi	Doka	Hanwa		Dan Mahawayi	Doka	Hanwa
Cultivated hectares:								
Gona	2.15	105	97	99	69.50	76	113	117
Fadama	0.45	107	91	107	330.43	77	121	109
Labor (man-hours):								
Family	1195	73	115	120	0.12	112	112	96
Hired	47	119	22	465	3.21	67	484	25

[a]The average value productivities of land and labor were estimated from the Cobb-Douglas function estimated for all the sample households in the study villages (Table 5.12).

TABLE 5.14
Determinants of Gross Return from Crops Per Hectare and Per Man-Hour, Zaria Area, 1966-67

Independent Variables		Gross Return from Crops per Cultivated Hectare (₦)		Gross Return from Crops per Man-Hour (₦)	
		b Value	Standard Error of b	b Value	Standard Error of b
Constant		289.65		1.1341	
Labor (man-hours per hectare):					
Family	X_1	0.2934*	0.0425	-0.0004*	0.0001
Hired	X_2	0.6524*	0.1605	-0.0003	0.0003
Land:					
Total cultivated (hectares)	X_3	-2.4139	7.2125	0.0378*	0.0154
Proportion cultivated that was fadama	X_4	539.2296*	192.7865	0.5816	0.4120
Other inputs (₦):					
Seed	X_5	1.0362	0.6926	0.0019	0.0015
Fertilizer (organic and inorganic)	X_6	2.7491	3.7541	0.0084	0.0080
Dummy variables ^{a}:					
V1	X_7	94.6522	50.8018	-0.0435	0.1086
V2	X_8	58.3441	48.6930	-0.0266	0.1041
R		0.6874*		0.5473*	
Syx		208.85		0.4464	

^{a}For definition of the dummy variables see Table 5.6.

When gross return was expressed in terms of the return per man-hour, an increase in family man-hour input resulted, as expected, in a decrease in the return per man-hour, while also, as expected, the number of cultivated hectares was positively related to the productivity of labor.

Although the signs on all the other variables in both models were consistent with expectations, none of them were significantly different from zero. The lack of significance of the seed and fertilizer input was perhaps not surprising because, as we stressed earlier, little in the way of improved technology had been adopted in the study villages at the time of the survey.[34] The significance of those inputs, however, likely will increase with the differential adoption of improved technologies.

Sales, Gifts, and Purchases

In the 1970-71 marketing survey, considerable attention was paid to major crop disposal patterns and the relationship of those patterns to other variables--such as overall levels of production, use of credit, quantity of grain in storage, timing of sales, market channel available, and prices. It is largely from this study, then, that our understanding of the role and practices of marketing emerges.[35]

Annual disposal patterns. About 24 percent of the millet production and 15 percent of the sorghum production were sold, with both production and amount marketed being greater for sorghum than for millet (Table 5.15). Of the fifty-four farming households in the sample, 80 percent sold some millet and 74 percent sold some sorghum during the year.

The table also reflects the non-market disposal activities, which also affect crop-disposal activities. As Smith (1962) explained, purely non-commercial transactions that are a complex set of exchanges derived from religion and kinship are an important part of the disposal picture. First, there are gift exchanges in set kinship contexts such as childbirth, naming, circumcision, marriage, and death. Second, Islamic tenets provide the context for transfers and exchanges at fixed festivals, such as Eid el Fitr and Eid el Kabir. Third, Islamic practice requires distribution of grain at the end of the fast (Ramadan or Azumi) and the transfer of grain-tithes at harvest (zakka). Religious alms (sadaka) are also distributed in expiation or propitiation. The gifts in set kinship contexts and for fixed festivals are usually both given and received by all households in the village, whereas gifts for zakka are mostly given by farmers for distribution to religious leaders, and poor, disabled, and elderly people in the villages. In the 1970-71 survey year all of those gift types accounted for 18 to 20 percent of the total foodgrain production, indicating the importance of social obligations. Farmers with the most production gave most in both absolute and relative terms, which tended to demonstrate status and affirm prestige in the community.

Sales and gifts together thus amounted to nearly 40 percent of production, whereas 60 percent was used for consumption within the farm household. That does not mean, however, that every household had a marketable surplus of grain and completely covered subsistence needs with 60 percent of produced grain.

TABLE 5.15
Average Household Production and Disposal of Millet and Sorghum, Zaria Villages, 1970-71

Variable Specification	Dan Mahawayi	Doka	Hanwa		Average of the Three Villages
			Non-cattle Owners	Cattle Owners	
Household size	10.8	8.6	11.6	13.8	10.7
Millet:					
Production (kg)	540	280	796	1756	699
Disposal (percent):					
Consumed	48	65	60	60	58
Sold	37	18	25	10	24
Gifts	17	17	15	25	18
Timing of sales--percent sold:					
Up to 6 months after harvest	48	48	28		41
Within 6 months of next harvest	52	52	72		59
Amount stored at harvest (kg)	438	239	694	1528	596
Percent in store after:					
1 month	82	84	73	89	82
6 months	45	56	48	63	52
11 months	23	17	10	34	20
Sorghum:					
Production (kg)	1235	1199	949	2562	1397
Disposal (percent):					
Consumed	58	68	71	64	65
Sold	21	17	4	10	15
Gifts	21	15	25	26	20
Timing of sales--percent sold:					
Up to 6 months after harvest	34	62	42		46
Within 6 months of next harvest	66	38	58		54
Amount stored at harvest (kg)	1046	1033	740	2185	1180
Percent in store after:					
1 month	92	99	84	88	92
6 months	65	44	42	50	52
11 months	18	12	12	14	14

A number of farming households reported purchases of both millet and sorghum within the year. Twenty-eight percent of the farming families purchased millet for their own consumption; 50 percent purchased some sorghum. In the aggregate, however, the quantity of grain purchased by all sample households in the three villages was less than the quantity of grain remaining in the stores at the beginning of the next harvest. Therefore, the marketable surplus amounted to at least what was actually sold. From a village standpoint, Doka had no real surplus, but there were surpluses in Hanwa and Dan Mahawayi. An examination of the sales distribution among the farming households showed that 19 percent of the farming families were responsible for half of the millet sold and 23 percent for half the sorghum sold.

Table 5.15 also summarizes the relationship between harvest and the times when farming families sold their foodgrain. Only 43 percent of the sales were at harvest or within six months of harvest; 57 percent were in the six-month period preceding the next harvest. Only about 2 percent of the foodgrains were actually sold at harvest.[36]

The data collected in fact provided an empirical basis for questioning many of the common assertions about farmers' marketing and storage situations and decisions, at least in the Zaria area.

One assertion often made about marketing in less-developed countries is that farmers are so desperate for money that they are forced to sell their crops at harvest. There was virtually no evidence that was true in the survey villages, at least with regard to foodgrains. Very little grain in fact was sold--more than 80 percent of the production went into the farmer's own storage facilities. Because most cotton and groundnuts had to be sold at harvest to the marketing boards for those crops--and the official purchasing season by the licensed buying agents is confined to a short period after harvest--that arrangement could have provided needed cash and encouraged farmers to hold off on grain sales.

Another assertion often made is that certain existing monopsony forces tend to exploit the farmers and deprive them of a fair share of the price paid by the final consumer. Monopsony forces are those which restrict the farmers' choice of buyer and/or the price the buyer pays. In such cases, the buyers have a substantial degree of control over the price, perhaps because of services produced for the farmer, perhaps because of other social or economic --patron-client, moneylender-borrower--relationships. Again, the Zaria evidence did not support this monopsonistic view. The study showed that grain producers had access to and used several types of buyers and a number of different market outlets. Nor did farmers appear to receive unfair prices; one farmer marketing grain at one place using a particular channel received prices similar to those obtained by other farmers using the same outlet. Only when farmers took their grain directly to the urban market did they receive higher prices--as they should have, because then they were performing more of the marketing services.

A third piece of common wisdom is that the marketing of crops in developing countries is tied to the extension of credit, which compels farmers to sell crops at inopportune times--when prices are low, for example--or under circumstances that contribute to a low price being paid. In the Zaria villages, however, extension of credit was associated with only a small percentage of marketings. In the few instances when that association was made, no explicit interest charge on the credit was involved in the transaction, nor was the farmer compelled to repay at a specific time. Most borrowing was from friends or relatives. In short, there was no evidence that marketing tied to credit was significant for sorghum or millet (Hays 1975a).

Finally, it is often said that farmers have inadequate or unsatisfactory storage and thus sell their unstorable surpluses at harvest to minimize losses -- but take a lower price to do so. Again, in the Zaria villages, grain-storage capacity was adequate in volume terms and it was relatively easy to expand the capacity through building another granary rumbu.[37] In addition, the cost of storage by using the rumbu was low relative to other possible techniques. Because cost-efficient and effective on-farm storage is so important to the farmers' abilities to store surplus grain throughout the year, we now examine storage practices in more detail.

Producers' storage practices. Giles (1965) identified six different storage methods in use among farmers in the northern Nigerian savanna; dried-earth granaries, granaries made of plant materials, underground stores, in-hut storage, clay pot storage, and, occasionally, modern silos. Only the dried-earth granaries, or rumbana, and in-hut storage facilities were found to be important in the Zaria villages. A rumbu is a specially built bin made from a mixture of dry grass and clay--somewhat like an oversized urn or pot. The 2.5 to 4.5-meter-high structure rests on large stones, to keep out rodents and to prevent the bottom from softening in the rains. It is covered with a removable thatch roof if the opening is on top.

Millet was stored toward the end of the rainy season (September) and sorghum in the dry season (December), so moisture content of either was low when placed in the rumbu. The relative dryness of stored material meant that storage losses in the Zaria area were considerably lower than grain losses in the southern region of Nigeria (Anthonio 1968). During our study, sorghum and millet were usually stored unthreshed, in bundles, and no modern insecticides were used to protect the grain.[38]

More than 85 percent of the farmers in the Zaria survey villages owned at least one rumbu. The remaining 15 percent of the farmers stored some commodities in the rooms of their compounds--what Giles (1965) called in-hut storage. The most frequent size of rumbu found in the study households was one which had a capacity of forty bundles of sorghum equivalent to 1.1 metric tons of threshed grain. Converted into terms of threshed sorghum, total household storage capacity, given an average of 2.6 rumbana plus the use of room storage, amounted to a farmer's storage capacity for grain of about 4.5 metric tons. The estimated annual storage cost per ton of grain stored by these methods was ₦1.00 (Hays 1975b). During the study year, storage capacity increased by more than 4 percent.

Of the total production of grain available at harvest, sample farmers in the survey villages sold approximately 2 percent, consumed about 4 percent, used about 11 percent as gifts, and stored about 84 percent (Hays 1975a). The 84 percent stored at harvest was then removed periodically for consumption, seed requirements, gifts, or sales. Table 5.15 shows the average quantity of millet and sorghum stored at harvest and the percentage remaining at different times after harvest as stocks were depleted.

Grain was removed from storage mostly for consumption needs, as reflected in the regular decrements to stored amounts. The need to obtain cash to meet certain expenses led to sales. The commonest reason for such sales was the need to purchase farm inputs in the June-August period. Most farmers stated, however, that it was important to "have in store more grain than would be consumed during the year in case of a bad harvest." Once it was determined that a normal harvest could be expected, a determination first made in July or August, extra grain would usually be sold to reduce stored amounts. That of course helped to depress prices of food grains further at harvest time.

The year 1969-1970 was a relatively normal year and eleven months after harvest -- that is, at the beginning of the new harvest--20 percent of the millet and 14 percent of the sorghum, on the average, remained in storage. According to the consumption study, the average household in these villages consumed 80 kg of sorghum a month. Specific consumption figures for millet were more difficult to estimate because the grain only partially went directly from storage to home consumption. A certain amount of grain trade and product manufacture (fura) took place before millet was recorded as consumed, so it was difficult to identify home production from reported consumption. However, by roughly estimating 32 kg per household per month, millet stores were sufficient to cover a slightly longer period.

THE LIVESTOCK PROCESS

In the farm-management study, we initially viewed livestock as household capital goods, which yielded quantifiable incomes in the form of offspring, various food products (eggs, milk, and meat), and services (particularly donkey transport). Animals were also recognized as contributing, in somewhat less easily quantifiable terms, to the maintenance of soil fertility. In the survey villages, small ruminants and poultry were often confined to the compound; droppings were swept up every day and periodically carried to nearby fields in baskets. The rotation of cattle on harvested fields in the dry season ensured that the organic manure was rationed out on selected fields. Many of the benefits were shared by the household, but quite clearly animal ownership itself was rarely collective. Within the household, animals were seen as the property of individuals. Thus, when one woman decided to break out of her marriage, she could and did sell her three sheep to pay for her flight from the village. The household benefit from her animals as sources of fertilizer was, of course, reduced.

Virtually all farm households included livestock at some time. In 1966-67, more than 90 percent reported owning some type of animals in the survey year. Sixty percent raised chickens, half raised goats, more than 40 percent had sheep, and 18 percent had donkeys. Fewer owned cattle (14 percent), guinea-fowl (8 percent), and horses (5 percent). Percentages of households owning animals in 1970-71 were generally reported to be even higher. Cattle ownership was, as mentioned earlier, confined to people identified ethnically as Fulani. Being Fulani, by contrast, did not ensure that the household owned cattle.

The concept of building up a herd of animals was clearly not operative; declines in inventory value were common in Dan Mahawayi and Doka. Even among Hanwa cattle owners a slight decline in cattle-holding value was noted during the farm-management study year. Though much of the decline in animal inventories was due to sales or special consumption requirements, on holidays, for example, other forces were also at work. In 1970-71, chickens almost disappeared from Hanwa at one time as a result of a mysterious epidemic. By the end of the expenditure-survey year, a few birds had been replaced, but the poultry population remained low. Because death not resulting from bleeding renders the animals unfit for human consumption, this broad decimation represented a substantial, although unmeasured, net loss of cash income from egg and meat sales, as well as a sudden depletion of peoples' savings accounts.

Cattle Ownership

If a similarly destructive disease, as that for poultry, had reduced the cattle population for Hanwa households, the impact on incomes and expenditures would have been more substantial; it would have been devastating because both the household economy and the cropping aspects of the farming systems in cattle-owning households were strongly linked to the operation of the livestock enterprise.

A few of the characteristics that distinguish cattle-owning households from those owning no cattle--hereafter referred to as non-cattle owners--have already been mentioned and shown in different tables. Cattle owners had larger farms (Table 5.2), so their claims to more land were perhaps facilitated by their relatively higher cash incomes to purchase the usufructuary rights, as well as by the fact that the village leadership was also Fulani and owned cattle. Cattle owners grew fewer crops on their land-- most in two-crop mixtures--and used significantly more hired labor (Table 5.3) and organic manure.[39] Cattle owners also generally had larger households (Table 5.2), but they devoted fewer man-hours per cultivated hectare per year to the crop operations and more man-days per year to other occupations, particularly herding, than did non-cattle owners (Table 5.5).[40] Higher cash-expenditure levels of cattle owners reflected higher incomes, as well as generally greater wealth--as measured by livestock and other capital goods.[41]

The animals owned by Hanwa cattle owners were basically unimproved white Fulani, which is a Zebu type. The majority of herds were adult cows more than three years old. Each cow had calved for the first time at three or four years of age, and thereafter calves were born every eighteen months--with each cow producing four to six offspring in productive years. Milk production was estimated to be only about 400 kg per lactation, although that output level varied widely. The low yields were presumably partly because the cattle were unimproved strains and were kept at poor nutritional levels. In the Zaria area, great reliance was placed on bush grazing, supplemented by limited amounts of crop residues immediately after harvest and, later in the dry season, occasionally excess cotton seed distributed by the marketing board after removal of the lint. Dependency on bush grazing contributed to considerable seasonal fluctuation in the weight of animals and therefore in milk yields.[42]

Though 24 percent of the animals in the Hanwa herds were bulls or bullocks in 1966-67, none of the male animals were used for field work or transport.

Hanwa cattle owners represented a transitional stage between Fulani transhumance practices and a sedentary mixed farming operation. While all households maintained permanent residences in the village, the herding family members had a somewhat separate existence in the bush for much of the year, as they followed their cattle, grazing along a path from 1 to perhaps 15 km away from the village. The structure and daily routines of the household reflected this dual existence. Both women's and older children's duties became more economically important and time-consuming.

The most striking contrast of cattle-owning with non-cattle-owning households was the relatively greater employment of older children, seven to fourteen years old, in the herding operation and their lesser involvement in crop-related work. Children in Hanwa cattle-owning households worked 25 days in fields on crop activities, while their counterparts in non-cattle-owning households put in 43 days on crops. But children in cattle-owning households spent 123 days a year in herding activities, while children in non-cattle-owning households spent only 3 days in off-farm work over a comparable stretch of time.

Our data on women's work in cattle-owning and non-cattle-owning households cannot be so clearly contrasted in terms of work-time allocations; women's responsibilities for processing and selling the milk in cattle-owning households, however, accounted for a great deal of time spent outside the compound. When the cows were within a few kilometers of the village, women often walked out at least once a day to help with the milking and to collect the milk. Other times the herders brought the milk into the village in the evening and ate a meal in the compound before returning to the herd.

Women's actual processing of the milk (nono) and butter (man shanu) each day generally took less than an hour, but if they also made fura (soured millet balls) for sale with the nono, as the majority in Hanwa did, an average of eight hours a day was needed to produce and sell the commodities in Zaria. This included walking time to Zaria but not the time to collect the milk from the herd. In contrast, Hanwa women pursuing other common occupations--cowpea-based commodities and the manufacture of groundnut oil and presscake (kulikuli)--required only five or six hours a day to do their work, done on a less regular basis. Only the two non-Fulani women who made tuwo for sale to workers in Zaria expended time on off-farm occupations on a scale similar to that of the Fulani fura da nono makers.

Because no animals were used in the tillage operations, complementarities between the crop and livestock enterprises were restricted to two areas: first, the provision of manure in exchange for forage from the crop stalks and leaf residues; and second, the availability of cash from the sale of livestock products to support larger crop operations than would otherwise be possible.

Evidence that Hanwa cattle owners' farms benefited from the first complementarity was found in the yield data.[43] Evidence that the second was true was confounded by the congruence between the village leadership roles as traditional land allocators and as livestock owners themselves, but in that tenure in Hanwa was so mobile, it would appear that the larger farm sizes accumulated by Fulani cattle owners was to some extent correlated with their better cash and wealth positions. Further, the cattle owners' greater hiring of farm labor during seasonal shortages also seemed to support the relationship.

Sales and Gifts

Ultimately it is the sale of milk that makes cattle-owning such a distinctive influence on household life. Milk is a "cash crop" that provides a steady income to supplement the production of other crops for consumption. And for Hanwa households, Zaria city's concentration of consumers provided a steady source of demand for milk. So reliable was this demand that quality changes through the year were tolerated. When supplies of milk were seasonally low, Hanwa milk processors diluted the nono with water and kuka--cream of tartar found in baobab tree pods--extending the supplies of milk to meet the demand of regular customers. Owners of the cattle themselves, it should be noted, had nothing to do with this practice. Men in the households owned cows, but the milk literally belonged to the wife or wives of the owner. A man with two wives, for example, was obligated to divide the milk his cows produced equitably between the wives. If one chose to sell it unprocessed as fresh milk (madara), that was her business. She was responsible for buying, with that money, the food she would prepare on the nights she cooked; but so long as she found her return sufficient, she could handle the milk as she wished. If the other wife chose to sour the milk, remove the butter, and sell the soured nono with fura, again that was her business.

Gifts of milk and butter to relatives and strangers were made, particularly for naming ceremonies and the like, but they appeared not to be significant in terms of volume. These products, unlike grains, did not appear to be used as zakka at any time. Even consumption of milk within the milk-producing households in Hanwa was more than twice that of non-cattle-owning households. But on average, members of Dan Mahawayi households--many of whom identified themselves as Fulani but all of whom purchased their milk from nomadic cattle-herding Fulani in the neighborhood--consumed amounts of milk about equal to the Hanwa mean.

THE OFF-FARM EMPLOYMENT PROCESS

Until recently, it was assumed in the literature that farmers engaged in tropical agriculture were, or aspired to be, full-time farmers. In rainfed areas, the dry season was assumed to represent a time of surplus labor and gross underemployment unless the farmers had access to irrigable land. The strategic implications of excess labor supplies and an idle season were, therefore, to encourage the development of irrigation opportunities to even out the seasonality of rainfed cropping activities and to employ available labor more fully. However, as we discussed earlier, empirical observation of farming systems currently practiced in the Zaria area suggested that both the assumptions and the implied strategies could be in error. Many nonfarm[44] jobs were available to, and taken by, Zaria farmers to provide additional sources of income.

In all three villages, the primary occupation of men was farming--both in 1966-67 and in 1970-71. But only 25 percent of the household heads in 1970-71 said they pursued no secondary occupation and many in fact had more than one such occupation. One farmer, in Dan Mahawayi, for example, traded a variety of crops and commodities on his private account as well as acted as a licensed buying agent for the commodity-marketing boards, owned the grinding engine in the village, and arranged transport services on occasion by leasing vehicles. He also provided loans and fulfilled certain official village government functions. It is likely that he also owned cattle and had put them out on loan (riko) with nomadic Fulani, although that ownership was not verified.

Types of Off-Farm Opportunities

As we indicated earlier, the location of the village appeared to be important in determining the level and composition of off-farm employment opportunities. It is therefore not surprising that the nonfarm opportunities existing in the two more distant Zaria study villages tended to be linked to the agroecology of the area and to the farming operation, whereas in Hanwa, the easy commute to Zaria city opened up a whole array of urban occupations as well.

Jobs of a wide range in various manufacturing, trading, and service activities were linked to the agricultural and ecological environment of the area. Mat-making, brown sugar manufacture, and calabash decoration were representative of the local manufacturing sector; local trading of crops, the hawking of various foods, such as roasted meat and bean cakes, and the sale of kolanuts were typical of the trade sector; donkey transport, building construction, and groundnut decortication were some of the regular service activities. Nearly all agroecologically linked jobs were also sex-linked in some way. Only men manufactured brown sugar (mazar kwaila); only women produced locust bean cakes (daddawa). Only men provided donkey-transport services; only women hand-pounded grain. Many job opportunities were further linked in some way to income or wealth status. Only poor people begged or offered head transportation services; lower-income people tended to repair bicycles. Koranic teaching implied higher income as well as social status; livestock traders also tended to have high incomes. Kolanut trading, cap embroidery, and well-digging seemed to be less linked to income status.

Few of these agroecologically linked jobs required extensive formal training—the major exception perhaps being Koranic teaching—and/or any steady commitment of time. Entrepreneurial skills, however, were at a premium. For most off-farm jobs, an individual had to identify and pursue the work opportunities single-handedly. Even hired farm labor work provided some room for wage bargaining; those offering their time as laborers needed to have some independent sense of the market and some prior notion of the time that would be required to complete a task. This was especially true of jinga workers, who normally agreed in advance with prospective employers on their expected payment.

Jobs that were less linked to the rural economy and were thus generally of more recent origin included wage labor outside of agriculture, tailoring with a sewing machine, and ownership or

operation of motor transport. Only residents of Hanwa had any
selection of these jobs, if they also concurrently wanted to farm.
Several Dan Mahawayi residents left farming altogether and moved to
Zaria or Kaduna to seek urban jobs full-time. But 32 percent of the
survey households in Hanwa in 1970-71 managed to farm and to have at
least one male adult pursuing full-time wage employment for regular
income as well.

Household income status in relation to these newer job types
was less easily defined than in relation to the rural occupations
that people had been pursuing for years. While gardening for a
university professor's household might seem menial enough to
classify as a low-income occupation by anyone's standard, the
regular wages and professional independence of the work might in
fact place a wage-gardener in an income bracket slightly above the
village average.

Jobs having no traditional roots were open primarily to men or
boys only. Women's exclusion was partly cultural, partly ascribable
to women's lack of educational opportunities, and partly related to
hiring practices used by Zaria employers. The practice of purdah or
auren kulle was perhaps the major cultural barrier to women's
employment in urban occupations, but the belief systems about
appropriate relations between men and women were perhaps more
fundamental. Many women in Hanwa, for example, often spent five
hours or more each day in Zaria as independent entrepreneurs sitting
on a street corner selling their largely male customers fura and
nono. Those same women would be reluctant, however, to seek or
accept a job that meant that they would sit in a factory supervised
by men for the same period of time. The educational question is
similarly colored by concepts of appropriate behavior. Although
both boys and girls could be trained in Koranic schools, many fewer
girls than boys were allowed to attend. The skill-learning
opportunities in more secular areas were similarly split; while boys
were often able to gain the fundamental ability to write their names
and apply for employment, girls rarely possessed the means to learn
even this basic qualification. Finally, hiring practices of
employers were often related to western stereotypes of appropriate
candidates as well as to the local standards--so some confusion and
flexibility reigned. Thus, cooks and cleaners--the sort of
low-skilled, low-paid jobs open to village farmer/job seekers--were
usually men; yet baby nurses were always women. Secretaries and
clerks were of both sexes; factory labor--only one Hanwa man--was
also often mixed, although supervisory levels tended to be male.

In contrast to the agroecological non-farm occupations, most of
the newer occupational types demanded more regular commitment of
time from those who wanted to pursue them. This distinction--rather
than an occupation's "traditional" or "modern" nature--might, in
fact, be the major difference between the two types of work.
Whereas village crop traders in business for themselves could work
regularly or steadily two or three days a week throughout the year
if they so choose, a clerk in a modern shoe store had no such
choice. Women preparing groundnuts for local sale could work or
not, depending on the price of inputs as well as competing household
demands, but a young women thinking about taking a job as a baby

nurse for an urban family would have no such freedom of time allocation.

Income Class and Off-Farm Work

The amount and composition of male adult off-farm employment, as noted earlier, was found to be influenced not only by the seasonality of agriculture but also by the ease of accessibility to Zaria. However, within villages it was apparent that the amount and composition of non-farm work was influenced by income class.[45]

Unfortunately, the significance of the off-farm employment component was not initially recognized at the beginning of the farm-management survey in 1966-67. Consequently, the data set obtained was not so complete as would be desirable for undertaking a comprehensive analysis. In addition, a much more detailed study relating off-farm employment to income class was undertaken several years later in the southern Kano province by Matlon (1979). Therefore, rather than attempt to draw conclusions from our own data, we defer most of the discussion on this subject to the next

TABLE 5.16
Work and Income Composition by Income Class, Zaria Area, 1966-67[a]

Variable	Per Capita Net Disposable Income		
	Low	Middle	High
Household income (₦/year)	132	232	300
Percent from:			
Farm	86	82	77
Off-Farm[b]	14	18	23
Household:			
Number of members	10.8	9.6	6.0
Number of male adults	2.3	2.5	1.9
Dependents/male adult	4.6	3.8	3.2
Hectares/household	3.3	4.5	4.3
Work on-farm (man-days/household):			
By household members[c]	314(95)	391(86)	319(91)
Hired labor	44	75	91
Work off-farm (man-days/household)[b]	83	162	222
Days worked/male adult:[d]			
Farm	128(0.37)	137(0.49)	156(0.74)
Off-farm	40(0.23)	58(0.26)	118(0.31)
Total	168	195	274

[a]Income class excludes taxes and income earned from cattle. The boundaries of the income classes were ₦5.2-₦16.1/capita/year, ₦16.2-₦33.4/capita/year, and ₦34.7-₦78.1/capita/year. Households included were those involved in both the farm management and consumption surveys.
[b]Figures exclude contribution by family female adults.
[c]Figures in parentheses represent the percentage contribution by family male adults.
[d]Figures in parentheses represent the return in ₦/man-day.

chapter, in which we look in some detail at Matlon's results. Before doing so, however, we can draw a few conclusions from our own data. For example, although traders and laborers were found in all income classes, craftsmen such as tailors or blacksmiths were confined to the high-income class, as were top village officials. The lower levels of skills and capital required for off-farm occupations that were undertaken by individuals in the low-income class resulted in their deriving the lowest returns per man-day worked in off-farm occupations. In addition to the differentials in return, incomes earned by male adults in the high-income classes were much higher because the amount of time each male adult worked in off-farm occupations was also much higher (Table 5.16).

Because households in the lowest income groups were larger and had more dependent members per male adult than did households in the higher income categories, the household picture is somewhat different from that for individuals. Low-income male adults earned the lowest wages off-farm as well as on-farm and generally had smaller farms to start with. But the low-earning status translated into even lower household income because of the relatively fewer earning members per household. Nonfarm earnings of such households were only a quarter of those derived by high-income households and farm incomes were less than half of farm incomes in high-income households. Nonfarm incomes accounted for only 14 percent of the total income in the low-income group, as opposed to 23 percent in the highest. The relative emphasis that middle-income households allocated to farm work was not reflected in significantly higher returns to farming for those households. Off-farm incomes contributed proportionally more to disposable incomes of middle-income households than to those of low-income households.

Women and Off-Farm Work

The contribution of women to the incomes of households in northern Nigeria has tended to be neglected in most village studies.[46] Because the farm management study in 1966-67 yielded incomplete information on the off-farm contribution of women, we made special efforts to rectify that omission in the expenditure survey undertaken in 1970-73.

The participation of women in the nonfarm sector was not so directly visible as that of men, and it was somewhat more difficult to quantify without time-allocation studies. Because women acted as independent entrepreneurs--primarily in the food-processing industry--they wove their unpaid domestic tasks in and around the activities that constituted the paid work they did. The incomes they earned from their businesses were kept separate from those of their husbands and were generally spent on somewhat separate categories of consumer goods as well. Women were expected to provide personal items such as soap, cosmetics, and cigarettes, dowry items for daughters and gifts for friends, midday food for their children and themselves, and personal travel. In households where husbands were unable to provide the goods and services expected of them, however, women's incomes from their off-farm work often appeared to compensate.

Women's off-farm employment in the Zaria villages can be

characterized by several attributes: choice of occupation, independence, participation, domestic work competition, and credit and gifts. We now briefly examine each of these attributes.

TABLE 5.17
Occupations of Women in Two Zaria Villages, 1971-73

Occupations	Dan Mahawayi 1971-72	Hanwa 1971-72	1973
Number of women in sample [a]	35	47	271
Food processing [a]	35	60	179
Food-processing services [b]	4	12	46
Crafts [c]	31	1	18
Trading	6	10	65
Medicine [d]	0	0	2
Number of occupations per woman	2.2	1.7	1.1

[a] More than fifteen products were produced by village food processors on a regular basis. The women frequently referred to the production of each line as a separate occupation.
[b] Food-processing services indicates those food-processing activities in which the processor did not possess title to the goods produced but merely performed a processing function on contract for another food processor or household cook.
[c] Crafts done in the villages were weaving, spinning, and a little embroidery.
[d] Medicine here specified only those women publicly producing identifiable products. In fact, many women performed midwifery functions for fees, but on a more occasional basis. Others were engaged in more clandestine forms of spiritual or herbal medicine for which they also received remuneration, but not all of those women could be identified and none was willing to give information.

Occupational choice. Women learned occupational skills by observation, heard about new opportunities through husbands and friends, and pursued various enterprises even when they clearly perceived that rates of return were declining. These factors both contributed to and emphasized the fact that Zaria women faced a limited range of occupational options. The structure of occupations reported by the women included in the sample expenditure survey in Hanwa and Dan Mahawayi in 1971 and 1972 and in a complete census of all women in Hanwa in 1973 is presented in Table 5.17. Food processing stands out as dominant. Commercial food-processing activities as conducted at the village level required only regular household equipment for the most part and, of course, most women readily learned the skill involved as they grew up. Competition was thus keen and switching of product lines was frequent. Spinning of cotton thread was still the major craft in Dan Mahawayi, although women in Hanwa reported that they no longer did spinning but did weaving and embroidery instead. Returns to spinning were estimated to be about ₦0.10 per month--less than that for any other male or female occupation. Yet women, when questioned as to why they continued to pursue so unprofitable an occupation, often cited it as something they could do and valued even the little amount of cash they were thus able to earn.

Independence. Women lived in close quarters with other women and shared domestic tasks routinely. Yet no women in any of the three villages believed that it was desirable or appropriate to cooperate with another woman in the conduct of an off-farm occupation. Even though in Hanwa women frequently lived in the same compound with other women making the same product (fura), no purchases of ingredients or preparation tasks were shared. The ability to draw on one's children's labor, however, was critical both to the entrepreneurial independence and to the choice of occupation. Not having an appropriately aged child to run errands and do the selling meant that a women might have to go out of business altogether or change lines of work--from fried bean cakes (kosai) to weaving, for example--even though it was known that profit margins would be negatively affected.

Participation. Women's identification of themselves as doing non-farm work--having some sana'a--was virtually universal; women's participation in such work was intermittent and highly variable. One work pattern is illustrated by the case of a woman who reported at the beginning of one interview series that she regularly performed two different food-processing occupations and one service occupation. The monthly interview returns given for this woman (Table 5.18) were typical of others, in that some commercial activity was undertaken in every month except the one she spent away visiting. Yet, in no month were all three stated occupations performed simultaneously.

To deal with that variation, a method of employment scores was devised.[47] Scores were calculated for each woman in an extended sample of occupational types in two villages--Hanwa and Dan Mahawayi--for two consecutive years. Based on the scoring technique, it was apparent that Hanwa women worked at their non-farm jobs with slightly greater regularity than did Dan Mahawayi women (Table 5.19). The Hanwa women scored an average of 64 on employment in their stated occupations, whereas Dan Mahawayi women scored 55. That was consistent with the fact that the two major food-processing occupations in Hanwa received employment scores of 78 and 89. Deflating possible returns to various occupations--calculated as the total number of possible work-days times the daily return--by these observed participation rates, the average return per occupation in 1972-73 was ₦3.40 per month in Hanwa and only ₦0.77 per month in Dan Mahawayi. Hanwa women thus earned an average of ₦6.00 per month in off-farm work, compared with ₦1.66 per month by those in Dan Mahawayi. This differential might account for the slightly greater participation of Hanwa women in work activities, demonstrated by their slightly higher employment scores. That Dan Mahawayi women pursued their commercial activities as vigorously as they did, despite such relatively low returns, may indicate the importance of this type of off-farm work and independent financial resources to rural women.

TABLE 5.18
Work Pattern in Different Occupations by One Woman, Hanwa, 1971-72

Interview Period	Occupation Number		
	One	Two	Three
1971: January	No	Yes	No
February	No	No	Yes
March	Yes	No	Yes
April	........Visiting out of town.........		
1972: January	Yes	"Stopped, no gain"	No
February	No	No	Yes
March	Yes	No	Yes
April	Yes	No	Yes

Domestic work competition. Women's domestic work participation complicated evaluation of their paid work, but domestic work needs did not often compete with the ability of women to pursue their business activities. Domestic work was shared among women and, because of auren kulle, some tasks identified as women's work in other cultures were performed by men or boys in Zaria, food shopping and laundry, for example. Only Fulani women in Hanwa did household shopping on any major scale. Routine cooking tasks, which were time-consuming as "cooking" implied pounding the grain to flour as well, were also shared by women in the household, following a more-or-less fixed rotation of responsibility. In about half the consumption-survey households, each woman cooked for two consecutive days, the frequency of her turn depending on the total number of women in the household. In an average-sized household, each woman might cook only two nights out of six.

One domestic task that appeared to require more time for women in households in Hanwa than in Dan Mahawayi or Doka was that of taking sick children to the clinic in Zaria--a long walk and a long wait in the out-patient line. In Doka, the men still took the children to the clinic in a town 8 km away if necessary; in Dan Mahawayi, clinic visits by anyone were rare, as the nearest clinic was even farther away.

Credit and gifts. Credit and gifts could facilitate or constrict women's ability to pursue a business successfully. Often women received their working capital (jari) to start a particular enterprise as a gift or loan from a husband, brother, father, or other male relative. When poor business decisions or simply poor business reduced the supply of working capital below the minimal amount, many of the women interviewed said that they would return to some occupation on the basis of upcoming occasions at which they

TABLE 5.19
Work Patterns and Productivity of Rural Women, Dan Mahawayi and Hanwa, January-April, 1971 and 1972

Occupation[a]	Dan Mahawayi			Hanwa		
	Nos. Women Employed	Employment Score	Net Return (₦/month)	Nos. Women Employed	Employment Score	Net Return (₦/month)
Food processing:						
Millet balls	7	50	0.95	11	78	5.52
Soured milk	0	-	-	19	89	7.51
Koko/kunu[b]	4	50	1.12	5	53	1.19
Fried bean cakes	3	46	3.15	6	63	4.31
Roasted groundnuts	2	25	0.11	3	38	0.16
Cooked cassava	0	-	-	8	39	1.04
Fried groundnut cakes	12	47	1.35	2	63	1.81
Services:						
Pounding for others	2	31	0.47	12	55	0.83
Crafts:						
Weaving	8	64	0.54	1	88	0.74
Spinning	23	60	0.10	0	-	-
Trading:	6	71	0.47[c]	10	47	0.31[c]
Total sample:	35	55	0.77[c]	47	64	3.40[c]

[a]Only occupations where data were collected from at least five different women have been enumerated by name in the table.
[b]Drinks made from sorghum and millet.
[c]Since each woman pursued more than one occupation (Table 5.15) the monthly return per woman was ₦1.66 in Dan Mahawayi and ₦6.00 in Hanwa.

expected to receive gifts. Though adashe, or revolving credit societies, were known, they were not common. Women often gave credit to customers--the fura makers and some of the groundnut oil manufacturers in particular--and, unless they were good managers or had honest customers, women sometimes found these credit obligations a heavy burden in maintaining a profitable operation.

Women food processors also sometimes found themselves in a conflict between business interests and their domestic roles when it came to gifts. If visitors arrived just as the day's production was ready for sale, the producer was torn between her obligations as a hostess and the quick loss of working capital. If she gave too little, she risked her reputation. Normally, a small portion of output appeared to be allocated regularly as gifts or for home consumption, but the profit margin for certain snack commodities was very narrow. We were unable, for example, to document a single case of dan wake manufacture that resulted in a clear profit!

Because of the difficulties in understanding the dynamics of women's nonfarm employment and in calculating profits and losses, women's incomes from their off-farm work were not recorded for a period consistent with any of the three surveys. Some estimates of the contributions that women's incomes could make to "household incomes"[48] can be made, however. By applying information regarding average returns to various enterprises and average participation rates of women in Dan Mahawayi and Hanwa, we highlighted the possible role of women's earnings in non-farm occupations to supplement men's incomes (Table 5.20). Their relative contribution was particularly important in low and middle-income households, less so in high-income households.

The importance of milk-processing as a source of income is stressed in the Hanwa figures. Most of the Fulani households were in the low- and middle-income strata according to the male-income-based classification, which however excluded income derived from cattle. When we explicitly estimated women's incomes from the milk--and the fura often manufactured for sale with the milk--the incomes of the average households in the middle-income group rose so significantly that they exceeded those of the high income group.

Including women's earnings in an income calculation was clearly important. Indeed, the expenditures that flowed from women's purchasing power helped account for a substantial portion of the apparent difference between incomes reported in 1966-67 and expenditures reported in 1970-71.

CHANGE IN ZARIA FARMING SYSTEMS

Characteristics of Zaria Farming Systems

The preceding analysis has illustrated both the uniqueness and common characteristics of farm households in the Zaria area. It highlights the difficulties associated with distinguishing household welfare from production success and failure, and with understanding the implications of household decisions for promoting agricultural change. Given the interaction of the technical and human elements with existing technology, most farm households adopted a risk-averse

TABLE 5.20
Women's Earnings and Hypothesized "Household Incomes" by Income Class, Dan Mahawayi and Hanwa, 1970-71[a]

Variable	Dan Mahawayi			Hanwa		
	Low Income	Middle Income	High Income	Low Income	Middle Income	High Income
Average number:						
Residents	6.9	9.4	8.6	12.5	12.3	8.0
Income-earning female adults	1.9	2.8	3.3	2.7	3.6	2.3
Male adults[b]	1.8	1.6	2.0	2.5	2.9	1.8
Estimated annual earnings:						
Percentage composition:						
Net farm income	49.6	66.0	64.3	42.3	48.5	55.7
Off-farm: male adults	17.1	18.5	19.8	5.3	9.4	25.7
Off-farm: female adults	33.3	15.5	15.9	52.4	42.1	18.6
"Household" (N)	191	402	513	402	616	501

[a]The breakdown into the various income groups is based on the classification used for Table 5.16 for 1966-67. Because the figures for net farm income and off-farm income derived by male adults were not measured in 1970-71, they were calculated by using the 1966-67 figures for net farm income and off-farm income per male adult and inflating them by the average annual inflation rate approximating 10 percent. In the case of the per male adult figure, the resulting figure was multiplied by the number of male adults in the household in 1970-71.

[b]In 1970-71 they were defined as being at least 20 years old, compared with a minimum of 15 years in 1966-67.

production strategy with a goal of ensuring minimum food security. Operating within that framework, farm households attempted to increase their welfare through: first, more intensified use of existing resources; second, occasionally combining existing resources with some form of improved technology; or third, increasing efforts devoted to off-farm employment.

The Effect of Change Upon Household Productivity and Welfare

Changes in technical and exogenous factors of importance to Zaria farming households are likely to be: increasing population growth; increasing access, particularly for Dan Mahawayi; changes in community norms, as individuals tend to pursue their own interests outside of family and community ties; and declining soil fertility. Where population growth is accompanied by better market access and, as in Hanwa, good sources of organic fertilizer, household welfare can benefit by increased productivity in both crop enterprises and nonfarm employment.

Analysis of individual productivity, however, strengthens the impression that income disparities are likely to grow over time, as persons with better access to resources use them more effectively. Hired labor can help to relieve family labor constraints in the labor bottleneck. The availability of fadama combined with good access to markets and knowledge of new crop technologies can, as in Doka, permit more intensive exploitation of resources. Individuals from households with higher levels of per capita income worked more days to earn those incomes, but their returns per man-day were also higher, reinforcing incentives to undertake further employment.

In general, nearly all households in the Zaria villages stand to benefit from the development of improved agricultural technologies, particularly for food crops. The ability of Zaria farming families to exploit current resources more productively will, however, depend on changes made in the socio-economic environment in which they exist. New resources and opportunities, rather than a reallocation of present resources, or more intensive work at current jobs, will be required to boost productivity and welfare.

Implications for Promoting Agricultural Change

The diagnostic phase of activity which has been reported in this chapter indicates both the potentials and problems likely to be encountered in efforts to promote agricultural growth in northern Nigeria. The need to recognize farmers' aversion to risk argues for an incremental approach to change; the need to recognize farming households' food needs argues for a strengthened focus on agricultural research. The difficulties of communicating new inputs and new information on technologies through the existing support and service delivery mechanisms is also emphasized. The lack of evidence that these mechanisms are now of any benefit to Zaria households is striking; on the other hand, evidence of a reasonably well-functioning private sector marketing system indicates that only marginal changes are necessary to enhance the role of that sector in promoting agricultural change.

NOTES

1. Some of the results from these studies are presented in Chapter 6.
2. They were 38 percent in Dan Mahawayi, 29 percent in Doka, and 43 percent in Hanwa.
3. On occasion, however, that was quickly rectified when enumerators married local women!
4. To achieve a high degree of accuracy, single-point registered types of data can be collected at infrequent intervals (Norman 1973b). A significant event occurring at one point in time would be an example of the former; an event not significant and occurring daily would exemplify the latter. Distinguishing data in this manner suggests the possibility of two levels of sample: a large one in which sampling errors are minimized and single-point registered types of data are collected, and a small one for frequent interviews from which both types of data are collected. Matlon (1977) successfully applied this approach in a later study in northern Nigeria. We would seriously consider this approach if we were to undertake the studies again.
5. Methodological details of the various studies undertaken are extensively discussed elsewhere (Hays 1975a; Norman 1967, 1972, 1973b, 1977; Simmons 1976a, 1976b, 1976c). Other references that provide some information complementary to the approaches used in our studies include Dillon and Hardaker (1980), Kearl (1976), and Connell and Lipton (1977).
6. The significance of this was underscored by our discussion of mixed cropping in Chapter 3.
7. Since we do not have long-term, time-series data for the Zaria villages, it is not possible to examine the assertion in this chapter. However, we develop this theme further in the next chapter by comparing areas in the West African savanna that differ in population density and accessibility.
8. Because of that less than 3 percent of the land was fallowed in 1966-67.
9. The average sizes of fields in 1966-67 in Dan Mahawayi, Doka, and Hanwa were 0.7, 0.6, and 0.5 hectares, respectively. In earlier work we calculated a fragmentation index for each farming family which expressed the distance of the fields farmed by each household, both from the place of residence and from each other. Although the index had some shortcomings it did indicate that the actual level of fragmentation in the three villages was less than it theoretically could have been (Norman 1967).
10. The comparison, however, is complicated by differences in the composition of nonfamily labor in the three villages (Table 5.3). For example, virtually free communal labor (gaya) was most important among the cattle-owning Fulani in Hanwa; the significance of contract labor (jinga) was greatest in Dan Mahawayi, and on the average commanded a wage rate--when expressed in per man-hour terms--47 percent higher than for work paid by the hour (kwadago).
11. The marginal value products of nonfamily labor (in N per man-hour) at the arithmetic mean levels of the inputs used in Hanwa and Dan Mahawayi in 1966-67 were N0.032 and N0.039, respectively, as estimated from the Cobb-Douglas function discussed later in the

chapter (Table 5.12). The wage rates per man-hour were NO.045 in Hanwa and NO.053 in Dan Mahawayi.

12. The productivity of individuals will not only be determined by age and sex but also by the task being performed (Hall 1970; Cleave 1974). Therefore, not surprisingly much controversy exists in the literature over how to compare different types of labor (Collinson 1972). Usually, some sort of weighting system is used to express them in terms of some common denominator such as man-days and man-hours. The weights we used were as follows: young child (under 7 years old) = 0.00 of a male-adult equivalent; older child (ages 7 to 14) = 0.50 of a male-adult equivalent; female adult (more than 14 years old) = 0.75 of a male-adult equivalent; and a male adult (more than 14 years old) = 1.00 of a male-adult equivalent.

13. Although the results were significantly different from zero at the 5 percent level when expressed on a total cultivated hectare basis, it is important to note this significance disappeared when it was expressed on an area-cultivated per-capita or resident basis.

14. Longhurst (1980) and Matlon (1979) working more recently in neighboring areas looked at the life-cycle issue more closely.

15. In the Zaria villages in 1966-67, the partial correlation coefficients between number of family male adults and farm size, and between male adults and cultivated area, both after being standardized for village location, were 0.4712 and 0.4953, respectively. Both were significantly different from zero at the 5 percent level.

16. The partial correlation coefficient between total disposable income and total number of man-hours worked by family members was 0.7071 when corrected for village location. The coefficient was significantly different from zero at the 5 percent level.

17. One criticism of our analysis, which may have contributed to our not getting more significant results, was that the models involved using data from the farm-management and consumption studies that were undertaken at different times. In doing so, we assumed that 1966-67 data on production and work time were a proxy for such data in 1970-71, when the consumption data were collected. We are not sure, however, that it was such a bad assumption because in looking at the millet and sorghum production figures derived from the 1970-72 marketing study and the 1966-67 farm-management study, we found a correlation coefficient of 0.8075--which was significantly different from zero at the 5 percent level--between the two sets of data. Also, for households in both studies, we found that they produced an average of 1,955 kg (190 kg per capita) of sorghum and millet in 1966-67 and 2,036 kg (199 kg per capita) in 1970-71. Since the technologies of producing these crops did not change between the two studies and the weather conditions were not very different, it is likely that the amount of time involved in their production was also similar.

18. The reason for expressing time worked by family members in total and not per-capita terms was to avoid multicollinearity problems through including disposable income per capita as well. The reason for including both variables was to permit differences to be expressed in the productivity of labor in different occupations.

19. For example, no significant relationship was found to exist

between the use of hired labor and per-capita calorie intake. Because of income constraints, particularly on a seasonal basis, lower than desirable per-capita calorie intakes are not likely to be compensated by greater use of hired labor vis-a-vis family labor.

20. Because individuals possessed only usufructuary rights to land, land was considered as a component of durable capital investment.

21. The significance of hired labor is further underlined by a study by King (1976a) in neighboring areas in which he found that an average of 74 percent of the credit borrowed under informal loans was used for hiring labor.

22. The functions estimated were as follows:

$$Y_1 = 2.81 - 0.40\log X_1 + 0.02X_2 + 0.01X_3 + 0.18X_4 + 0.11X_5$$
$$\qquad\quad (0.04) \qquad (0.14) \qquad (0.01) \quad (0.04) \quad (0.04)$$

$$R = 0.72*$$
$$S_{yx} = 0.16$$

$$Y_2 = 2.74 - 0.52\log X_1 + 0.04X_2 + 0.01X_3 + 0.16X_4 + 0.19X_5$$
$$\qquad\quad (0.06) \qquad (0.18) \qquad (0.01) \quad (0.06) \quad (0.05)$$

$$R = 0.70$$
$$S_{yx} = 0.21$$

Where:

Y_1 = Total man-hours per cultivated hectare
Y_2 = Family man-hours per cultivated hectare
X_1 = Number of cultivated hectares
X_2 = Proportion of cultivated land that was fadama
X_3 = Organic manure (metric tons/hectare)
X_4 = Hanwa = 1; others = 0
X_5 = Doka = 1; others = 0
* = Significantly different from zero at the 5 percent level

23. Regression analysis reported elsewhere (Norman 1972), with hired man-hours as the dependent variable, indicated that in addition to this relationship, there was a significant negative relationship between the amount of hired labor and size of household. Perhaps most interesting about the results was the positive relationship between the use of hired labor and off-farm employment of family members, which included time spent tending livestock, working on farms of other households, and other occupations not necessarily directly connected with agriculture. Off-farm occupations provided a means of obtaining cash; it is likely, then, that the time devoted to other occupations by household members acted as a proxy for such earnings. Thus, an increase in time worked off-farm would indicate a household's greater ability, all other things being equal, to hire labor. Results of the study of expenditures confirmed this conclusion: farm-labor expenditures were highly correlated with total expenditure, a proxy for income, with an elasticity of expenditure exceeded only by that for clothing (Simmons 1976c).

24. Value of home-produced food was imputed on the basis of consumer purchase prices for comparable quantities.

25. The groundnut oil that was produced in the villages appeared to be used by the women who specialized in producing

certain cooked foods for sale and for which it was a preferred ingredient in terms of taste; in all these products, substantial amounts of groundnut oil were used. For soup, however, palm oil was certainly a cheaper substitute; it was also said to be preferred for its color and taste in the soup.

26. In contrast, in the much harsher climatic environment around Sokoto, millet and sorghum from rainfed land constitute both food and cash crops.

27. Replacing household size with a dependency ratio--that is, number of dependents per male adult--yielded similar results.

28. The partial correlation coefficients--controlling for village location--between food production and household size and between food production per capita and household size were 0.5688 and -0.0860, respectively. Only the first one was significantly different from zero at the 5 percent level.

29. Various other models using proportions of time spent at other activities yielded similar unsatisfactory results.

30. This was not unexpected, because after controlling for village location, a partial correlation coefficient of 0.5060 was found between work on the farm and size of the farm. This was significantly different from zero at the 5 percent level.

31. The functions estimated were as follows:

$$Y = 0.36 + 0.01X_1 + 0.14X_3 + 0.07X_4$$
$$\qquad\quad (0.006)\quad (0.04)\quad (0.04)$$

$$R = 0.49*$$
$$S_{yx} = 0.17$$

$$Y = 0.29 - 0.12X_2^2 + 0.34X_2 - 0.10X_3 + 0.08X_4$$
$$\qquad\quad (0.06)\quad (0.12)\quad (0.04)\quad (0.04)$$

$$R = 0.50*$$
$$S_{yx} = 0.17$$

Where:

Y = Proportion of value of production of crops that was sold

X_1 = Cultivated land (hectares)

X_2 = Cultivated land per resident (hectares/resident)

X_3 = Hanwa = 1; others = 0

X_4 = Doka = 1; others = 0

$*$ = Significantly different from zero at the 5 percent level.

32. In analysis presented elsewhere, we showed that farming families in general were allocating resources to crop production in a manner consistent with the goal of profit maximization (Norman, Pryor, and Gibbs 1979)--thus indicating allocative efficiency. However, we have not done so here, because we recognize that the validity of the conventional approach to testing this can be questioned since it is unlikely that any one farming family used all its resources at the general mean levels.

33. Obviously, all other things being equal, the average value productivities of the inputs are influenced by the degree to which they are used. The results for the different villages in fact generally reflected that the basic characteristics of gona, fadama, non-family, and family labor were similar in the different villages

and that location of the villages did not have a major influence on the productivity of those inputs used. In other words, variables V1 and V2 in Table 5.12 were not very important in determining the productivity of resources that were actually used.

34. It is not surprising that earlier analysis showed significant results when net farm income per cultivated hectare and per man-hour was simply estimated in terms of family labor input on the farm while controlling for village location (Norman, Pryor, and Gibbs 1979).

35. Attempts were made in the 1966-67 farm-management survey to obtain some information on marketing activities, but for various reasons the results were not considered to be accurate (Norman 1972).

36. Gilberts' (1969) study in the Kano area also found that farmers held grain surpluses for sale six months before the new harvest.

37. The plural is rumbana.

38. Bungudu (1970) found in a village almost 200 km from Zaria, local plant materials being used for protection purposes.

39. In Hanwa for the 1966-67 survey year, the total amount of manure and the rate of application for cattle owners was 25.71 metric tons and 7.12 metric tons per cultivated hectare, while for those not owning cattle the equivalent levels were 9.22 metric tons and 4.30 metric tons per cultivated hectare.

40. In Hanwa for the 1966-67 survey year, the input per cultivated hectare was 675 man-hours for cattle owners and 1,129 man-hours for those not owning cattle. The latter, however, had more labor-demanding fadama land.

41. In Hanwa for the 1966-67 survey year, the average household figures for cash costs incurred in crop production and total disposable income were N33.97 and N567.63 for cattle owners, and N21.60 and N221.30 for those not owning cattle.

42. Raay and Leeuw (1974) studied seasonal changes in fodder availability. The significance of improved nutrition and breeding on milk yields is underscored by work at the Shika station, formerly part of IAR, where 900 kg per lactation is commonly obtained from purebred white Fulani and 1,760 kg from crossbred white Fulani/Friesian (IAR 1971).

43. For example, for a millet/sorghum mixture in Hanwa in 1966-67, the average values of production per hectare and per man-hour input on fields farmed by non-cattle owners were N70.96 and N0.08, respectively. Comparable figures for families owning cattle were N80.22 and N0.13.

44. We use the terms off-farm and nonfarm interchangeably.

45. This was defined on the basis of per-capita rather than household income.

46. A major reason for this of course includes the practice of women seclusion, which makes it difficult to include them in conventional farm-management surveys, for which male enumerators usually are employed. Also, there may be a bias toward male chauvinism in such studies undertaken by males!

47. If a woman worked at all occupations that she identified as hers in the relevant time period before an interview, she received a score of 100. If she reported that she had done some, but not

164

others, she received a score that indicated the percentage of time
that she could have participated in a particular occupation that she
actually did. The woman whose work pattern is given in Table 5.18,
for example, reported that she had actually worked at her three
occupations ten times out of a possible twenty-four. She received a
score of 42. Employment scores were also calculated for the
occupations included in the sample. A high employment score for an
occupation suggests that it was regularly pursued; occupations with
lower scores were performed more sporadically.

48. Because there was no pooling of incomes by individual
earners in the household, this term has to be used cautiously. A
"household income" is a strictly theoretical concept. Since male
household heads had the responsibility of supporting their
households and controlled the most resources to do so, their incomes
in one sense constituted "household incomes".

6
Diversity
and the Context for Change
in Farming Systems

"Principles of actions can be set out, but the application
of these principles must take into account the different
geographical and geological conditions in different areas,
and also the local variations in the basically similar
traditional structures."

Nyerere (1971)

In the preceding chapter we examined in some detail the farming
systems practiced by farming families within three Zaria villages
during a particular time period. Such specificity in both location
and time, however, tends to mask the dynamic interactions
responsible for producing the farming systems currently found. The
diversity existing in farming systems across the West African
savanna not only reflects current interaction between the technical
and human elements but also reflects, to differing degrees, what has
happened in the past. In the same way the farming systems of the
future will be partly a function of what is happening now.[1]
Therefore, farming systems tend to be both location- and
time-specific. It is possible to realize, and appreciate, the
diversity existing in farming systems in the savanna only by
broadening the geographic scope of our discussion beyond the three
Zaria villages and by examining changes that have occurred over
time. Such an exercise can, in a general way, improve our
understanding of the different ways the technical and human elements
interact, as well as give us some idea of general trends in the
farming systems found in the savanna and the general types of
problems that will need to be addressed if the welfare of the
region's farming families is to be improved.

Material from many studies throughout the West African savanna
provides the data base for this chapter although the detailed
presentation of empirical data has been confined to studies with
which we were closely associated in northern Nigeria. Because of
differing objectives and methodologies of many studies cited, we
have had some difficulty in combining the results in a comparative
analytical framework. Some of our conclusions therefore should be
considered indicative rather than definitive in nature. In
addition, to demonstrate complexity of the interaction of the
technical and human elements, we depart in this chapter from the

more systematic presentation of the various components of the farming system used in Chapter 5. After reviewing the underlying significance of the technical element in partially explaining the diversity of farming systems in the savanna, we look at the impact of the human element and the changes that, as a result, continue to take place in the farming systems in the region. This naturally leads into a discussion of the distribution of resources and welfare. We then close the chapter by discussing the implications for bringing about constructive changes in the welfare of West African savanna farming families in the future.

THE DIVERSIFYING INFLUENCE OF THE TECHNICAL ELEMENT

Unlike certain aspects of the human element, variations in the technical element are not closely aligned to political boundaries but have a more regional distribution. As we have emphasized in Chapter 3, water is a critical ingredient in the farming systems found in the savanna. The degree to which water is available is undoubtedly a primary determinant in differentiating the farming systems, affecting particularly the allocation of resources such as labor.

Combination of Processes

In Chapter 3 we indicated that the amount of rainfall in the West African savanna not only decreases as one moves northward, but also is accompanied by an increased variability at the beginning and end of the rainy season. The progressively shorter growing season which results is paralleled by changes in the significance of and relationships among the crop, livestock, and off-farm components of farming systems.

The types of crops that can be grown on rainfed, or gona, land are more limited as one goes north, with the cropping systems based on mixed sorghum and millet in the Northern Guinea and Sudan ecological zones giving way to those dominated by millet in the Southern Sahel (Table 3.1). At the same time, cotton and groundnuts, major export cash crops, also disappear from the cropping systems.

Table 6.1 illustrates the types of adjustment that take place when two specific crop mixtures are grown in different ecological areas in northern Nigeria. We observed that in the drier area of Sokoto (Table 3.2), the average number of plant stands per hectare was much lower than in Zaria, reflecting the farmers' response to the poorer soil-moisture expectations in the area.[2] The lower number of stands per hectare found in the Sokoto area consisted of a much higher proportion of millet stands compared with other constituents in the mixture, reflecting the comparative advantage that millet enjoys in the drier areas. The yield per stand of millet was much higher in the drier area, and yields per stand of other crops were correspondingly lower.[3] As a result, a higher yield of millet per hectare was obtained in the Sokoto area than in Zaria, whereas the yields of other crop constituents were correspondingly lower. In total, the overall value of production per hectare was lower in the Sokoto area than in the Zaria area,

TABLE 6.1
Two Mixed Cropping Enterprises, Northern Nigeria, 1966-68[a]

Variable	Millet/Sorghum		Millet/Sorghum/Cowpeas	
	Sokoto	Zaria	Sokoto	Zaria
Man-hours/hectare[b]	505.1	611.1	558.5	734.4
Numbers of stands/hectare	10,626	22,506	16,272	28,260
Ratio of millet to other stands	1.0:0.9	1.0:2.0	1.0:0.5:0.4	1.0:2.0:1.0
Yield (kg/stand):				
Millet	0.16	0.05	0.09	0.05
Sorghum	0.04	0.05	0.03	0.05
Cowpeas	-	-	0.02	0.02
Yield (kg/ha):				
Millet	892	370	772	400
Sorghum	186	768	·124	714
Cowpeas	-	-	63	167
Value of production (N) per:				
Hectare	49.94	66.05	45.26	76.33
Annual man-hour[b]	0.11	0.12	0.13	0.13

[a]The results for Sokoto refer to 1967-68 and those for Zaria to 1966-67. The same applies to other tables in this chapter where data from both areas are presented. Data for Bauchi were also collected in 1967-68.
[b]Excluded time travelling to and from fields and for threshing.

TABLE 6.2
Productivity of Upland and Lowland, Sokoto and Zaria, 1966-68[a]

Variable Specification	Upland		Lowland	
	Sokoto	Zaria	Sokoto	Zaria
Man-hours/hectare	484	540	1,042	1,298
Number of stands/hectare	17,710	30,648	b	72,970
Value of production per (N):				
Hectare	40.54	55.60	105.67	180.35
Man-hour	1.20	1.12	1.15	2.27

[a]The system used in calculating the entries in the table involved weighting the different enterprises according to their relative areal contribution to upland and lowland that was cultivated (Norman 1972).
[b]Not available.

although the returns per man-hour were similar, partly because the man-hour input per hectare was lower in the Sokoto area.

Cropping systems on lowland or fadama are, of course, somewhat less dependent on rainfall patterns. More readily available water, combined with higher-quality soil means that--as we stressed earlier--lowland is potentially more productive than upland (Table 6.2). Thus availability of lowland can have a significant impact in diversifying cropping systems, even to the extent of substituting additional crop activities for livestock and off-farm employment in contributions to family welfare.

However, since in most of the West African savanna there are only limited areas of lowland, the significance of livestock, particularly cattle, increases from south to north. That is a natural trend because the increasing shortage of water progressively reduces the potential for intensive cultivation systems while at the same time increasing the comparative advantage of extensive grazing systems. This trend is further strengthened in drier areas by the diminished threat of the tsetse fly, the carrier of the protozoan disease trypanosomiasis. Therefore, currently the potential for crop/livestock interaction is greater in the drier than in the wetter areas. That potential for a symbiotic relationship is reduced, but by no means precluded, because livestock, particularly cattle, are often in the hands of nomadic Fulani--sometimes called Fulbe, Peulh, or Fula--throughout the region. The migratory pattern of these herders, which we described in Chapter 3, has a number of advantages, among them the possibility of using the northern drier areas unsuitable for crop cultivation during the rainy season, and during the dry season enabling manure to be produced for crop cultivators in the more southern areas in return for the use of crop residues. The Fulani, in addition to owning cattle themselves, also herd cattle owned by crop cultivating families.[4] Despite the apparent dichotomy between the day-to-day management of livestock and the cultivation of crops the traditional symbiotic relationship between livestock herders and crop farmers--in which crop residues for the livestock and manure for the fields are important elements--has generally worked well in areas with relatively low population densities. Such relationships are also well adapted to the ring cultivation system.

The third major component of the farming system, off-farm employment, is, unlike crop and livestock activities, less dependent on rainfall. Therefore, all other things being equal, it would be reasonable to hypothesize that off-farm employment would become relatively more important in the farming system as the length of the growing season decreases; that is, as one moves northward through the savanna. Although we suggested earlier that the location of the village is an important determinant of both level and composition of off-farm employment, some members of farming families in drier areas practice another strategy to overcome the problem of being in a village located unfavorably in terms of opportunities for off-farm employment: they migrate for a short term, seasonally (Ravault 1964; Roch 1976; Goddard 1971). Such short seasonal migration tends to be concentrated in the dry season and tends to involve males between the ages of fifteen and forty-four (Sutter 1977; Faulkingham 1977). Usually, this causes few problems, although Faulkingham (1977) in Niger reported that onions grown in the dry season had to be cultivated by younger family members because of the departure of older males during this period. The types of jobs, which vary enormously, include cutting and selling firewood, pushing hand carts, and helping to harvest crops in areas farther south, for example, cocoa in Ghana and the Ivory Coast (Beals and Menzies 1970; Faulkingham 1977). One particular industry in theory complementary with rainfed agriculture is the tourist industry in Gambia. The height of the tourist season is during the dry season; consequently, many individuals in farming families engage in work connected with tourism (Peil 1977). There is a problem, however, with some off-farm occupations, particularly those such as the stranger-farmer system in Gambia and Senegal that involve migration during the rainy season, and result in increasing the dependent-to-worker ratio back at home. Haswell (1975) has noted that increasingly middle-age males are migrating during the rainy season, thereby depleting the productive labor force in agriculture in the home villages.

Seasonal migration can help the welfare of farming families in two ways: first, by contributing positively to the income of farming families; and second, perhaps just as significant in areas where food supplies can be precarious, by reducing food claims on home produced supplies in that the migrants feed themselves while away from home.

In accordance with the preceeding discussion on the reduced rainfall northward through the West African savanna, we can conclude that, all other things being equal, the significance of rainfed agriculture in determining the welfare of farming families is likely to decline relative to the significance of livestock and off-farm employment. However, the phrase, "all other things being equal," is important in lending validity to this conclusion. For example, cattle (as we showed in Chapter 5) require a significant capital investment, preventing some farming families in drier areas from owning them even though cattle have a natural comparative advantage compared with rainfed agriculture.[5]

Impact on Labor

The degree to which water is available has a major impact on the level of labor a farming family will allocate to crop activities, livestock activities, and other off-farm employment.

The figures in Table 6.3 show that family male adults, the main contributors to work on the family farm in northern Nigeria, allocated relatively fewer of their days worked to farm activities involving crops and relatively more to off-farm activities, including livestock, in the drier Sokoto area. But in addition to providing some verification for the shift in the relative significance of the different processes as one moves northward, study of Table 6.3 once again points to the critical issue of seasonal bottlenecks: the decreased length of the growing season in the drier Sokoto area in fact accentuates the seasonal labor bottleneck. With short dry season migration being a primary way to salvage low opportunity cost labor during the long dry season in Sokoto (Table 6.3; Figure 6.1), the seasonal allocation of labor between farm and off-farm activities becomes an important issue.

IMPACT OF THE HUMAN ELEMENT

Although facing similar characteristics in the technical element, farming families in different areas and even within specific areas may have different farming systems. We have already discussed extensively the importance of the human element as a determinant of this differentiation.

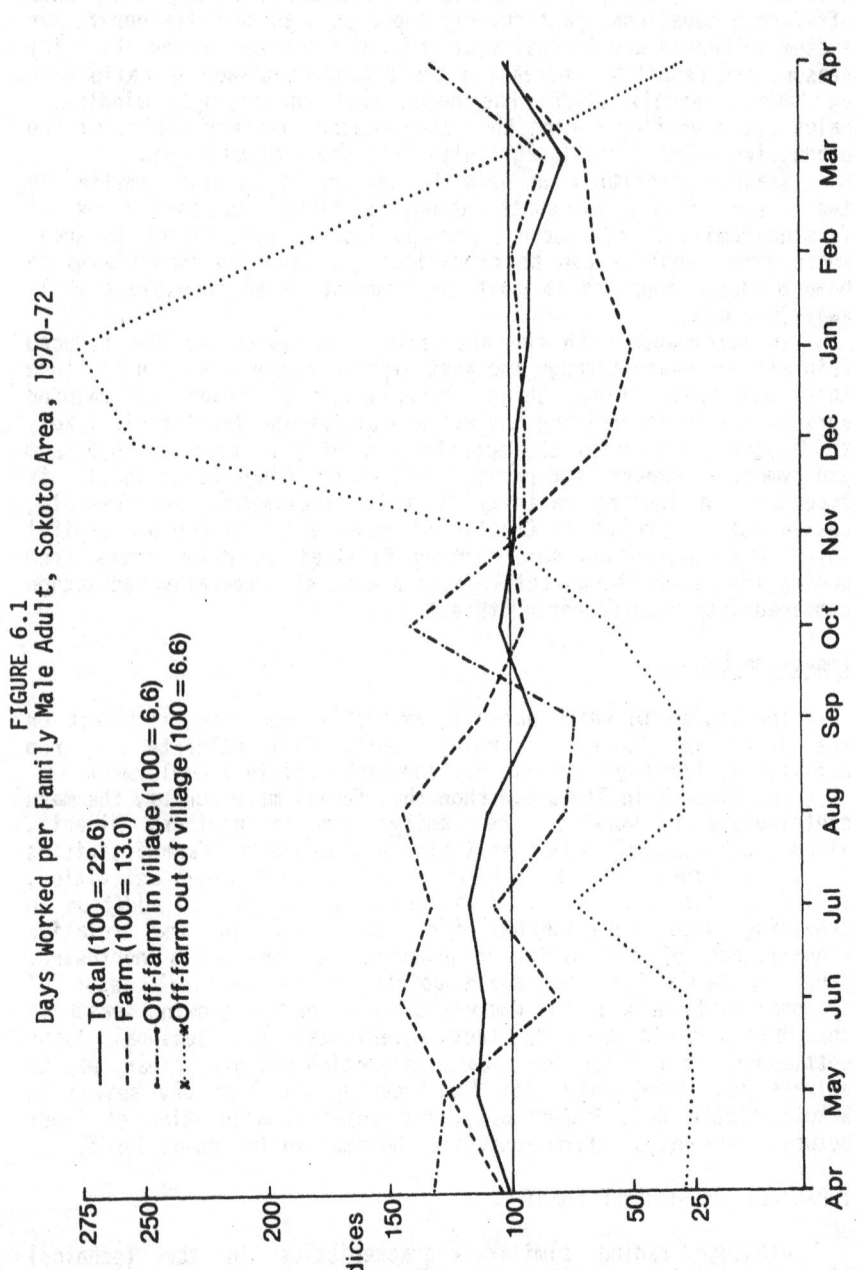

FIGURE 6.1
Days Worked per Family Male Adult, Sokoto Area, 1970-72

Total (100 = 22.6)
Farm (100 = 13.0)
Off-farm in village (100 = 6.6)
Off-farm out of village (100 = 6.6)

Indices

When we examine the impact of the human element, we have great difficulty in separating the relative significance of the current interaction of the technical and human elements from what has happened in the past. Therefore, in the following sections we combine the discussion of differences among and within areas with an examination of changes over time. The rationale for doing that is simply that many of the current differences in farming systems across the savanna have been partially determined by differences based on what happened in the past.

TABLE 6.3
Relationship Between Seasonality of Agriculture and Work, Northern Nigeria, 1966-68

Variable	Sokoto	Zaria	Bauchi
Work per male adult per year:			
Days	273	230	231
Percent breakdown:			
Farm	58.4	61.3	58.0
Off-farm: Village	28.7	38.7	42.0
Out of village	13.9	-	-
Work on family farm in average month:			
Total man-hours on family farm[a]	130.5	150.0	110.5
Hours per day worked by male adults on family farm[a]	5.0	4.4	4.7
Busy period for farm work			
Four busiest months:			
Months	June-Sept.	May-Aug.	June-Sept.
Percent of total man-hours on family farm	56.6	50.4	53.2
Peak month:			
Month	July	June	July
Total man-hours on family farm[a]	258	256	210
Family male adults:			
Hours per day worked on family farm[a]	6.1	5.0	5.3
Days: Farm	19.9	16.8	19.2
Off-farm	7.0	7.6	6.5
Total	26.9	24.4	25.7

[a]Excluded time travelling to and from fields.

We believe that two exogenous characteristics are important in differentiating current farming systems both among and within areas: communications and population density. Changes in these determinants over time directly or indirectly encourage or force

adjustments in the farming systems practiced by farming families.
An historical perspective is particularly important in understanding the development of communications. During pre-colonial days, many parts of the West African savanna had important communication linkages across the Sahara to North Africa. However, with the advent of colonialism--first Portuguese, -then British and French--communications became more oriented toward the coastal areas of West Africa, a trend accentuated by the construction of railways from the coast into the hinterland to facilitate the evacuation of export cash crops such as cotton and groundnuts. The orientation to the coastal areas continues to the present day, while communications within the savanna are still generally poorly developed.

Population densities have been increasing substantially in recent years. Population growth rates of the West African savanna countries averaged 2.5 percent from 1970-79 (World Bank 1981b).[6] All these countries are still basically agrarian, with an average of 73 percent of the labor force employed in agriculture in 1979 (World Bank 1981b). This continued concentration of the labor force affects the farming systems practiced by farming families, particularly in areas where population densities are high in relation to the carrying capacities of the land available for farming.

The interactive influence of changes in communication and population density, we believe, partially help explain the current diversity in farming systems, and also the changes that are occurring over time. Hence, these factors influence what we believe is a society in transition. To provide some structure to our discussion we examine the diversity and changes under three main headings: community norms and beliefs, external institutions, and resource ratios. In structuring our discussion, however, we emphasize that the diversity and changes in the farming systems are the result of interaction among all exogenous and endogenous factors.

COMMUNITY NORMS AND BELIEFS

The Village in a Traditional Setting

Traditionally, villages in the West African savanna, as we indicated in Chapter 3, generally have been characterized by a strong sense of community within given hierarchical systems of control (Remy 1977; Ramond, Fall and Diop 1976; Lewis 1978; Kohler 1968, 1971, 1972; Jones 1976). Many of these hierarchical systems were based on the inhabitants' longevity of residence in the village and on status at the time of arrival. For example, as pointed out by Jones (1976) and Haswell (1975), in many villages the founders, who had the leadership roles, were joined by more recent settlers and by those who originally came as slaves. In general, researchers do not consider that hierarchy to be very exploitive.[7] Three reasons commonly are given for this conclusion:

1. Communal land tenure systems characteristic of the region, combined with low population densities, have been the rule

(Hill 1972; Maymard 1974).
2. As pointed out by Haswell (1975), traditionally communities had a concept of shared poverty with poverty being determined primarily by the technical element (i.e., climate and soil)--creating in a sense a community "welfare state."
3. In the traditional savanna societies of West Africa, despite low levels of capital, the concepts of mutual obligations and of the gift have been very important (Mauss 1954; Vercambre 1974). One concept (mutual obligation) prevented large inequalities from developing in income distribution; the other (the gift) helped cement the social fabric and, through an ideology that stressed redistribution (Watts 1978; Hill 1972; Raynaut 1976), militated against accumulation of economic assets.

Based on substantial qualitative evidence, those characteristics generally describe the social fabric of the traditional villages throughout the region, except for minor differences due to variations in ethnic origin (Pelissier 1966), religion, and culture.

The Winds of Change

Changes associated with the development of improved communications within the savanna and rapidly increasing population densities are, we believe, contributing to adjustments in the community norms and beliefs within villages. Many of the changes taking place emanate from the increasing significance of strategies designed to create economic independence rather than the traditional strategies of preserving social and economic interdependence, a theme we discussed conceptually in Chapter 2. Such a breakdown in community solidarity is modifying and, in some cases, weakening social sanctions against behavior by individuals and families that in earlier times would have been frowned on or even forbidden because it went against the interests of the community as a whole. Therefore, while the potential for individual initiative has been increased, so also has the possibility, given particular situations, for the development of greater intra-community inequalities, founded not on social structure but on changes in the relationships of production.[8]

Many examples of the changes in community norms and beliefs are discussed in other parts of the book, including the way in which labor is hired, the decreasing amounts of food stored (Guggenheim 1978), the increasing individualization of land tenure, and increasing monetization of the economy (Monnier et al. 1974). However two deserving special attention are those influencing the behavior of community leaders and changes related to the structure of individual families.

Village leadership. In Chapter 4, we discussed an example of changes in community norms and beliefs that permitted the village leadership, if so inclined, to seize the limited amounts of lowland in order to reap the benefits of the highly profitable dry season tomato technology package. Weil (1970) cites an example in Gambia where the introduction of oxen resulted in the more influential families demanding the return of land they had previously rented to other families in the village. The increased potential for abuse of power by village leaders obviously needs to be taken into account in designing strategies for improving the welfare of the mass of farming families.[9]

Families. In Chapter 3, we suggested a few reasons that the complex family units, once so common, are increasingly being replaced by nuclear family units--a phenomenon happening throughout the West African savanna. This trend in family structure, in a sense symptomatic of the individualization that is taking place in the community as a whole, is influencing the progressive breakdown in the relationships associated with management of fields by traditional complex family units. Under the traditional system, fields farmed by families were divided into common and individual fields. The common fields, controlled by the family head, provided food for all members of the family. Now, more of those fields are coming under the control of other individuals in the family, resulting in a decrease in obligations of family members to work on the common fields; hence, there is no longer the assurance of food from the common farm to meet subsistence needs. Increased individualization of fields (Unité d'Evaluation 1978; Kleene 1976) has encouraged the growing of cash crops for the market.

The increased decentralization of decision-making within families, however, is also creating problems in introducing improved technology, especially where an input by the extension service or an institutional credit program is directed only at family heads. Niang (1978) reported that individuals other than the family head grew cotton in the Experimental Units of Senegal even though cotton was, in revenue terms, less profitable than some other crops. They did so because payment for the improved cotton inputs which were distributed was made at the end of the season by receiving a lower price per kilogram for the crop. Formal credit programs for the inputs of the more profitable groundnuts had to be channeled through the family heads, many of whom were unwilling to take responsibility for individuals working on their own fields. Venema (1978), also in Senegal, has given another example of problems involved in increased individualization of fields. His data for groundnuts indicated that both the average sowing date and the date of first weeding generally were later for fields under the control of individuals than for those under the family head. As a result, yields were also lower. That situation emphasized that a shortage of labor existed; and because family heads ensured that labor requirements on common fields were given priority, labor inputs were lower, and timing of operations was poorer, on individual fields. In addition, use of improved technology was lower for reasons mentioned above. There appear to be no easy solutions to the problem, unless the difficulties in directing credit and programs for distributing inputs to individuals rather than to family heads are addressed.

Thus, pressures for adult males to break away to form their own nuclear family units are likely to be strengthened by such experiences. This option is not open to women because no matter what the family structure is, they are in a dependency relationship with men when it comes to farming matters in the West African savanna. That does not mean that women in the region do not contribute significantly to agricultural activities and that changes are not occurring. The custom of secluding Moslem women prevalent in northern Nigeria is not characteristic of large parts of the savanna. But factors that have been found to be important in determining the level of agricultural activity by women are ethnic origin, type of task (Delgado 1978; Guissou 1977; Piault 1965), type of crop (Haswell 1953; Maymard 1974), and structure and size of the family (Milleville 1974). In other words, there appears to be a certain flexibility in the role of women in farm work. Also, acceptable norms of behavior seem to be permitting changes over time, although there is little evidence that they are benefiting women. In fact, the reverse often appears to be the case. Guissou (1977), for example, indicated that the increasing popularity of nuclear families, which tends to increase the dependent/worker ratio, means that many women have to work harder than under the old system. Haswell (1975) has noted that in Gambia the building of causeways into the river had encouraged women to switch from growing early millet and hungry rice (Digitaria exilis) to cultivating swamp rice, which involved harder work in a less healthy environment.

What adjustments then, in the light of these changes, are farming families making in terms of the goal(s) they pursue, and in their attitudes to risk and uncertainty? It is difficult to be definitive about goals but what evidence there is suggests that most farming families continue to favor food self-sufficiency as their major goal rather than relying on fully commercialized agricultural production where profit maximization is the major goal. However, a number of factors are encouraging greater marketing of crops, sometimes at the expense of food self-sufficiency. Examples include: increasing individualization of fields; the availability of improved technologies for growing export cash crops, such as groundnuts and cotton, though not for growing food crops; external pressures, such as support systems being skewed toward cash crops for export (Lele 1975); the compulsion and need for money to pay taxes, repay credit, or use otherwise (Campbell 1977; Nicolas 1960; Jones 1970; Lewis 1978); and the economic necessity of maximizing the return from a very limiting resource, such as land (Matlon 1977).

A recent study by Balcet and Candler (1981) in northern Nigeria suggested that the decision-making behavior pattern of farming families may be lexicographic, with the aim of food self-sufficiency dominating from the time the rains start until the first food crops germinate. If the rains are "good" this "food focussed" period will usually end at the first weeding; if the rains are "bad" it will extend much later into the season. After this critical stage is past and the farming family feels sure of its food supply, objectives slowly change to income maximization, as additional information is gained on what the year will be like. Therefore, what the farming family does during the second phase to fulfill the subsidiary income-maximization objective will depend to a great extent on decisions made earlier in the year to fulfill the food self-sufficiency objective.

Attitudes of farming families on risk and uncertainty are

important in determining their goal(s) and the types of improved technology they are likely to adopt. Indirect evidence shows that risk-aversion strategies are important for farming families in the savanna. There is, however, insufficient empirical information from other parts of the West African savanna to indicate whether farming households differ in their risk attitudes, in so significant degree to account for differences in farming systems.[10]

The risk-aversion strategies adopted by farming families can be divided into two broad classes: one having the objective of avoiding price variability and the other having the objective of minimizing variability in yield or production.

1. The objective of minimizing price variability can be recognized in the strategy used by many families to pursue a goal of at least some degree of food self-sufficiency, thereby reducing the risk of having to purchase food at considerable price variability in the market. Other examples include planting cash crops only after food crops have been well established (Jones 1976; Balcet and Candler 1981), thereby supporting the goal of food self-sufficiency, and the tendency during drought periods to decrease the production of cash crops in favor of food crops.

2. Strategies designed to minimize variability in yield or production are exemplified by the practice of growing crops in mixtures, sometimes consisting of different species but also on occasion of different varieties of the same species (Charlick 1974; Kohler 1971). Another strategy is to grow a number of crops rather than one or two, because not all crops are similarly affected by varying conditions in weather, insect infestation, and disease attacks. Yet another strategy is the traditional preference of the spatial scattering of fields, especially in areas of low-population density, to take advantage of micro-environmental variations (e.g., soil conditions, rainfall variations, disease attacks, etc.).

Although practice of the above mentioned strategies seemingly provides evidence that families are risk-averters in agriculture, the arguments are not conclusive. For many of the examples, there may be more than one reason why such strategies dominate. For example, we showed earlier that growing crops in mixtures not only is a risk-aversion strategy but also is consistent with the goal of obtaining a higher return per unit input of land and/or labor. Furthermore a family's practice of diversifying the number of crops grown is consistent with the notion of rotating crops to provide the potential for a more even use of labor throughout the year.

EXTERNAL INSTITUTIONS

Agricultural policies are country specific. Because government or government-linked agencies tend to dominate in providing the support systems serving agriculture in the West African savanna, a very close relationship exists between the support systems and the agricultural policies from which they evolve. The support systems therefore tend to be differentiated by national boundaries.

Export Cash Crop Bias of Support Systems

The current support systems still reflect, in various degrees, the systems that were developed under the French and British colonial administrations. Both colonial powers set up support systems that encouraged the production of export cash crops, particularly cotton and groundnuts, sometimes at the expense of domestic food production. In the francophone countries, a coordinated approach to the provision of support systems--such as extension, improved inputs, institutional credit, and product marketing services--often has been made possible by commodity development agencies.[11] The most successful projects, judged only from the perspective of the adoption of improved technologies developed for cotton and groundnuts, have been in the francophone countries. Evidence is found in the differences in the yields of cotton (for example, in Mali Sud) and groundnuts (for example, in the Sine Saloum in Senegal) and in the relative degree to which oxen have been successfully introduced in such francophone areas, compared with anglophone areas.

Even up to the present time, support tends to be concentrated in particular areas where success has been achieved in increasing export cash crop production, as in the Sudan and Northern Guinea ecological areas, where rainfall is more favorable for such crops. Hence, geographical variation in providing support systems is yet another factor contributing to regional differences in farming systems.

Even within areas where good support systems are present, the diversifying influence on farming systems can be important for those farming families who do make use of the support systems compared to those who do not. Some of the changes that take place are reviewed briefly in the following sections.

Levels of technology for food and cash crops. Support systems for export cash crops--both on the input and output side-- have enabled the improved technologies developed for their production to be adopted. The development of improved technologies for such crops was supported during colonial days, a trend which continued to some extent into post independence times.[12] Usually, improved technologies have been applied to export cash crops, while traditional technologies generally still are being used for producing food crops, for which improved technologies are still generally unavailable. Because of the research emphasis on growing crops in sole stands, the significance of sole cropping has increased in conjunction with the adoption of the improved technologies for export cash crops, particularly in the areas covered by commodity-development operations in the francophone countries. In contrast, in northern Nigeria, cash crops commonly are still grown in mixtures, although Stubbings (1978) noted, in one of the Agricultural Development Projects initiated with the World Bank's support, a trend towards increasing significance of sole crops. He speculated that trend may be because of the extension service encouraging farmers to plant crops in sole stands. The practice of sole cropping is even extending to food crops, particularly where strong support systems are found and animal traction has been successfully introduced,[13] although most food crops throughout the savanna are still grown in mixtures (Charreau 1978; Delgado 1978; Niger, Ministère de l'Economie Rurale 1973).

Changes in capital, credit, and cash. There is ample evidence that capital investment in agriculture throughout the savanna in West Africa traditionally has been low (Kafando 1972; Ernst 1976; Jones 1976). Apart from livestock, most of the capital has been produced with labor inputs during the dry season, when the opportunity cost of labor was low.

With the introduction of improved technology, the type of some of the capital used by farming families has changed significantly thereby creating the potential for greater diversification of farming systems. Purchased in the market place rather than being produced with labor at the village level, the new types of capital include among others, most types of animal-traction equipment, sprayers, and inorganic fertilizer. The use of such capital is likely to continue to increase as farming families adopt improved technology. That implies that it is necessary for farming families to enter further into the market economy in order to provide funds for paying for such capital. Also, with the introduction of improved technology, the level of capital investment required increases, as does the proportion of capital that has to be obtained through purchasing in the market place, in both relative and absolute terms.

A major problem with respect to the extra cash is that the seasonal cash flow tends to be inversely related to the level of agricultural activity (Dunsmore et al. 1976). The time when agricultural activity is approaching its peak, June to September, is the time of major demand for expenses in agriculture, but that coincides with the time that cash resources are at their lowest ebb (Matlon 1977). With the introduction of improved technology, the problem is likely, initially at least, to be exacerbated. Variations in the seasonal cash flow are worsened by the fact that the business of farming and the family itself are not separated. As a result, extra pressures may arise during the crop-growing season, if the food grain stores run low and supplies for feeding family members have to be obtained from other sources.[14] Savings and credit are obvious ways to overcome seasonal cash flow problems.

Traditionally savings often have been mobilized by reciprocal relationships among people and through the selling of livestock. Interestingly enough, unlike the higher rainfall areas of West Africa, references to traditional savings and credit clubs in the savanna are few. Bouman (1977) mentions their existence in Senegal while King (1976b) refers to their being present among women in northern Nigeria. The lack of coincidence in the savanna region between the expenditure and income cycles would appear to provide opportunities for introducing institutional savings programs, but unfortunately there has been almost no attempt to do that. A small program in northern Nigeria apparently had some success (Huizinga et al. 1978a and 1978b), and a bank in Mali would like to organize a similar savings program (Bank manager at Koutiala, 1978). It is unfortunate that, in general, emphasis has been solely on institutional credit programs rather than on programs that recognize the complementarity between savings and credit. Placing the emphasis solely on institutional credit, without a savings component, limits the potential for developing self-sustaining credit programs.

Traditionally, credit obtained from local sources has been used primarily for consumption purposes. In the light of the preceding discussion, that is not surprising. On the one hand, cash expenses in agriculture traditionally have been minimal; on the other, the need for food has been combined with substantial social obligations (e.g., naming ceremonies, marriage expenses, etc.) and the need to pay taxes. Both have contributed to a bias toward credit for consumption. One problem in analyzing credit from local sources is wide variability in interest rates. For example, studies in northern Nigeria revealed both high explicit interest rates (Vigo 1965) and low or even zero rates of interest (Matlon 1977; King 1976a). A number of factors, however, can obscure the real or implicit interest rates. Three of them are as follows:

1. Farmers in the region are likely to be reluctant to disclose not only loans they have received but also the interest rates they are being charged. The reluctance to disclose the latter is because usury is usually frowned upon in Islamic societies.
2. Loans that are given often involve reciprocal social obligations. For example, in return for a loan, an individual might be expected to work on the fields of the creditor at a time when his (the borrower's) labor has a high-opportunity cost. That in effect would be an interest payment, although it probably would not be articulated as such in the agreement drawn up for the loan. Therefore, it is unlikely that such obligations would ever be expressed explicitly as interest rates.
3. There is often a masking of the interest rate when loans are paid back in kind. For example, loans are sometimes repaid at harvest with in-kind payments when the prices of the harvested crop are lowest.

Increasingly, the potentially exploitive nature of traditional credit systems has been emphasized (Dubois 1975; Clough 1977), as community structures break down and individualism increases with the concomitant increase in contact with external institutions (Watts 1978; Raynaut 1976). Clough (1977) has discussed how intervillage wholesalers who store grain in rural areas use urban credit to secure large profits from seasonal price movements and also extract a flow of grain through harsh credit relationships with farmers in villages. As we indicated in Chapter 5, however, we were unable to confirm this in the Zaria villages study.

There appears, therefore, to be some confusion over the implicit or explicit interest rates charged for credit at the local level. It does appear, however, that the interest rate charged can be heavily influenced by its source, by the amount of collateral the borrower possesses, and by the position of the borrower in the society. For example, King (1976a) found in his study that a substantial amount of credit at the local level was borrowed from relatives or close friends. In such cases the interest rate appeared to be zero or minimal. Haswell (1975) noted that lower interest rates were charged on loans given to people who owned cattle. Perhaps related to this is an observation by Hopkins (1975) that farmers in Senegal preferred to borrow money rather than sell livestock as an emergency source of cash. Because farming families in the region as a whole possess only usufructuary rights to the land, in the eyes of the law such land cannot be used as collateral. However, at the local level, credit arrangements do sometimes involve the use of land. Pledging land in return for a loan is becoming increasingly common (Goddard 1972). Finally, the position of the person in the society also seems to have an influence on the interest rate charged. Clough (1977), for example, observed that a lower interest rate was charged on loans to wealthy and influential people than to those who were poorer.

Institutional credit programs have been implemented throughout the savanna areas of West Africa to encourage the adoption of improved technology. This credit, sometimes given in-kind, has been of two types: short-term (such as fertilizer or improved seeds) and medium-term (such as oxen equipment). Two criteria are often used in evaluating such credit: repayment rates and, perhaps less commonly, equitability of access.

Generally, high repayment rate levels are achieved only when such programs are carefully coordinated with other external institutions and support systems, particularly input distribution and marketing of the product (King 1976b; Belloncle 1968). Certainly that has been true with respect to the introduction of oxen draft systems. One of the major concerns in institutional credit programs is the so-called misuse of credit for consumption purposes. Consequently, in an attempt to prevent that, credit has often been given in-kind, not always successfully. Venema (1978) has given an example in Senegal of fertilizer obtained on credit being resold at two-thirds of its value to provide cash to meet urgent consumption needs. He has also cited cases in which groundnut seed obtained on credit was collected by rich farmers from their debtors. Cases have also been reported of medium-term credit being misused; for example, the selling of oxen equipment and animals. The concept of raising interest rates on institutional credit, as advocated in the AID Spring Review (USAID 1973), the better to cover transaction costs and to discourage misuse of credit either through selling the products or through on-lending to other farmers, has, insofar as we know, not been attempted in the savanna areas of West Africa.

With reference to equitability of access to institutional credit, cooperatives or "pre-cooperatives" often have been the basis of governmental policy to encourage equitability--for example in Senegal, Niger, Gambia, and Nigeria--but because of organizational problems in traditional village societies, success often has been limited (King 1976b; Gentil 1971a and b; Storm 1977).

Economics of animal traction. While in many Asian societies animal traction has been used for centuries, it has only a fifty year history in West Africa. Because of this the introduction of animal traction into the West African savanna is usually through the market place. But a farming family purchasing animals and/or the necessary equipment must generate a substantial surplus of farm income over and above what is required for family survival. Generally, surplus crop production must be sold in the market place. It is not surprising, therefore, that as we indicated earlier, widespread adoption of animal traction has been closely linked with the use of improved technologies for export cash crops. Animal traction rarely has been successfully adopted where only food crops are grown, possibly because of a lack of relevant improved technologies--to provide the potential for generating sufficient surplus production--combined with product-pricing policies that have favored people living in urban areas.[15] Consequently, any support systems developed for food crops, which until recently were not common, have tended to be ineffective.

The economics of animal traction, moreover, are increasingly being questioned even in areas where there is a profitable export cash crop to provide revenue for the animals and equipment. In recent years prices of cash crops have increased relatively less rapidly than those for animals and equipment (Steedman et al. 1976; Traore and Toure 1978), which not only is slowing down the adoption of animal draft power, thus preventing the beneficial interactive effect between crop production and livestock, but also is creating a danger of aggravating the dual economy that is developing between those farmers who do have oxen and equipment and those who do not. We address that topic later in the chapter.

Integrated Agricultural Development Projects

A more holistic view of farming systems and their multiple constraints to increased productivity is reflected in the implementation of development projects now emerging. In many countries, the commodity-orientated development projects are being transformed into Integrated Agricultural or Rural Development Projects. Although the value of such integration is increasingly being recognized, certain problems are associated with implementing such policies, including these three:

1. Changing the philosophy of the implementation agencies themselves from the one-commodity approach to that involving more components of the farming system is not easy to do.
2. The lack of relevant improved technologies for food crops and a continuation of pricing policies which inhibit their production, particularly of millet and sorghum, are likely to mean that emphasis, in the short term at least, will continue to be on cash crops.
3. Except for the introduction of animal traction, little attempt has been made to integrate crops and livestock, though the issue is being addressed in a project recently commenced in Gambia (USAID 1978).

The desirability of having a more holistic view of the development process is obvious in that it would allow exploitation of the complementary relationships among the various enterprises as well as in the support systems. For example, one possible way that suppport systems for cash crops could help to stimulate food production is by encouraging the use of oxen on food crops as well as cash crops, with credit being repaid by revenues derived from the cash-crop component. That is, in essence, what the World Bank, which is the prime mover behind many of the Integrated Rural Development Projects in Sub-Saharan Africa, is moving toward (World Bank 1981a). It is now advocating setting up no more projects with elaborate support systems in areas where relevant improved technologies do not exist. In the short-to-intermediate run, this means concentrating such projects in the more favorable areas where export cash crops can be grown. In other words, the World Bank explicitly recognizes that support systems without relevant improved technologies available to produce surplus production inevitably have little impact. At the same time it is advocating increased research activity to develop relevant improved technologies for those areas where none are currently available. That implies a need for greater allocation of research resources in the food crop area, a view with which--because of the African food crisis and on equity grounds--we heartily concur.

The way those issues are dealt with in real situations, of course, is closely linked to the development of agricultural policy, which is briefly considered in Chapter 8.

RESOURCE RATIOS

Land-extensive techniques such as shifting cultivation, where soil-fertility regeneration is achieved through fallowing, have dominated traditional farming systems in Africa. The ring cultivation system we discussed earlier is simply a variant of the land-extensive system. Under high ratios of land to labor, the rational strategy, if income maximization is the goal, is to maximize the return per unit of labor, particularly with respect to labor used during periods when labor is most in demand. When land is relatively abundant, and technologies which substitute for labor are not available, labor bottlenecks, as we discussed in Chapter 5, become a major issue. As population densities increase, however, land-to-labor ratios decrease and eventually land becomes a significant constraint. We now look at these two constraints in the context of the savanna agriculture.

Labor Bottlenecks

The type of seasonal labor bottleneck that exists in the West African savanna not only is a function of the rainfall distribution but also is determined in part by the type of technology employed, including the power base. At the risk of over simplifying we believe we can make the following generalizations on technology and seasonal bottlenecks.

1. We showed in Chapter 5 that with only hand labor and indigenous technology, the time and amount of weeding is often most limiting. The weeding bottleneck might be

183

accentuated if the rains are particularly good (Unité
d'Evaluation 1978). Land preparation and planting also are
sometimes considered to be bottlenecks, particularly when
timing is important. Timing becomes significant as one
moves northward and when the growing season becomes shorter
(Unité d'Evaluation 1978).

2. As will be seen later (Chapter 7), the introduction of
improved land-intensive technology (e.g., seed and
fertilizer), without changing the power source, shifts the
bottleneck to the time of harvesting the increased yields.
That statement, however, should be interpreted carefully
because timing is still a particularly critical factor in
the weeding operation (Haswell 1953; Matlon and Newman
1978).[16] Also, for certain crops, one can argue that time
of harvest is really not such a serious bottleneck because
the rains are over and further serious damage to crops in
the field is unlikely, although we recognize that maize can
be attacked by rodents and other cereal crops by birds.

3. Changing the power source from hand to animals, but
retaining indigenous or traditional technology, apart from
ridging equipment to be used with oxen,[17] only accentuates
the weeding bottleneck. Larger areas of land often are
prepared which, because weeding equipment is inadequate,
have to be weeded mainly by hand (Tiffen 1971; Jones 1976).
Also, the harvesting bottleneck becomes more accentuated
when land preparation and hence planting operations are
carried out more quickly and efficiently than before the
power change (Figure 6.2).

4. A combination of animal power with ridging, planting, and
weeding equipment together with improved land-intensive
technology eases the weeding bottleneck, but it tends to
accentuate the harvesting bottleneck even further (Faye
1978), although that can be eased somewhat by using a cart
for evacuating the harvest from the field.

Animal traction, therefore, has been perceived as a way of
overcoming labor bottlenecks through increasing the productivity of
labor, and yet its use does create new labor constraints. We
already have mentioned the economics of animal traction and
indicated that introduction has been most successful where it has
been combined with improved land-intensive technologies for export
cash crops. Other emerging problems, however, concern the
successful adoption of animal traction. Three of them are
implementation, maintenance, and utilization.

Implementation. Commonly, training animals is difficult,
especially if operators are inexperienced. For example, animals
often have to be retrained at the beginning of each year, and it
takes the operators two to three years to get used to handling them,
particularly for some of the more skilled operations such as
inter-row cultivation during weeding periods (Wilde 1967; Wilcock
1978). Another problem often mentioned is the difficulty which
farmers not owning potential draft animals have in acquiring the
finances to pay for them. Often credit for animals is not included

FIGURE 6.2
Seasonality in Time Worked on the Family Farm Using
Mainly Traditional Practices, Zaria Area, 1966-67

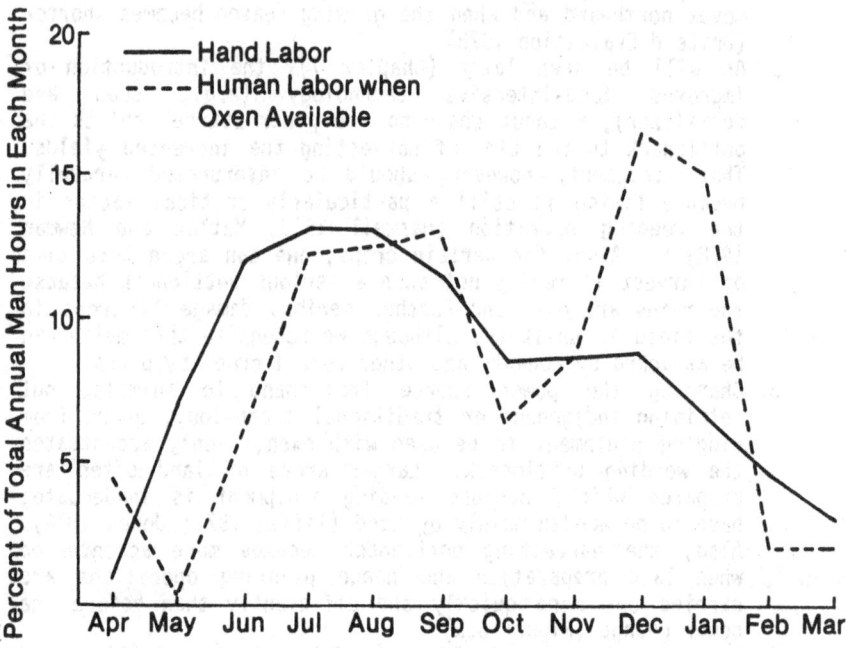

in institutional credit programs. In a survey undertaken in the Eastern ORD of Upper Volta, it was found that only one-fifth of the farmers could have obtained draft animals without the institutional credit program, which in that area included a draft-animal component (Barrett et al. 1978). It is perhaps significant to note that families in Mali owning animal traction equipment tended to have more cattle than did those not owning such equipment (Unité d'Evaluation 1978). Finally, equipment for animal traction often is not available on time (Jones 1976) or it is inadequate in terms of design--necessitating excessive repairs and other services (Dunsmore et al. 1976; Rocheteau 1975).

Maintenance. Often mentioned are problems of feeding and housing the animals, protecting their health, and finding available labor to maintain them (Venema 1978). Veterinary services often are poorly developed, and because finding sufficient food for the draft animals during the dry season is difficult, they commonly run with the cattle herds. As a result, they have to be retrained at the beginning of each year and also may be in poor condition. It has been recommended that they be fed supplemental feed grains at this time of the year to put them in good condition at the beginning of the rains, when their work loads are heaviest. But such recommendations are rarely followed (Weil 1970), which is not surprising in view of the opportunity cost of scarce grains during

185

this period. Problems are particularly acute in the southern part
of the region where N'Dama cattle are commonly used. These cattle,
although resistant to tsetse fly, are small in configuration and
therefore possess potentially less draft power than do the larger
animals found in most parts of the region.

Utilization. Efficient use of animal traction requires two
characteristics of the fields: that they are not too fragmented and
that they be destumped to prevent damage to the equipment. The
Experimental Units in Senegal have specifically addressed these
issues by giving financial incentives to farmers to destump their
fields and by encouraging their consolidation (Faye and Niang 1977).
Many times, however, these conditions are not fulfilled, resulting
in reduced work efficiency. Another problem, which we discussed
earlier, repeatedly is mentioned: the introduction of inappropriate
equipment in the sense that its use does not help overcome the most
pressing bottleneck. In Mali, Jones (1976) noted that farmers in
the late 1960s rarely used animal-drawn implements in weeding
operations. Though time involved in the plowing operation had been
reduced by half as a result of using the implements for that
purpose, it was not possible to double the area cultivated for
sorghum and millet because of the weeding bottleneck. As a result
the area cultivated increased by only one-fifth. The seasonal
variability in the tasks draft animals are expected to perform
constitute a major problem. Reference is sometimes made to the
potential significance of the cart as a way of increasing the use of
draft animals throughout the year (Zalla 1976). Yet this equipment
is not always available under institutional credit programs, despite
the fact that it can provide a significant source of income and is,
according to one study, the only part of the draft animal program in
Niger that has met with any real success (Charlick 1974). Using
draft animals to undertake operations on farms of other families is
another possibility, and according to Charlick (1974) citing
Nicolas, most farmers Nicolas surveyed in Niger used their equipment
primarily for rental purposes rather than on their own land.
Apparently renting out equipment in the francophone countries has
never been discouraged. In contrast, in northern Nigeria where
animal traction programs were initially implemented, farmers were
officially forbidden to do contract work for other families (Alkali
1969) because of government's fear that participating in such work
would have a negative impact on the condition of the animals.

Land Constraint

In traditional settings in West Africa characterized by low
population densities and indigenous technology, including a
hand-labor power base, the area cultivated has been closely
correlated with available labor and quality of the land (Kohler
1968, 1971). Rapidly increasing population densities, however, are
creating increasing pressures on the land resource, with at least
three disturbing effects.

Farm size decrease. As we discussed in Chapter 5, land in such
situations is rarely fallowed and generally is rented out or leased

if not in use by the owner. Such changes in land-tenure arrangements obviously are a response to the increasing monetary value of agricultural land as it becomes more scarce. The implication of that trend is serious in the sense that finding new ways of maintaining soil fertility becomes more and more urgent, as both the length of the fallow period and the amount of land fallowed are decreased. Indeed, the changes in the tenure relationships may be exacerbating the problem of deterioration in soil fertility. Luning (1963) observed that in northern Nigeria people renting land were discouraged from applying organic fertilizer, apparently because if they did they might try to retain control of the land. There have been reports of land being recalled by the landlord because of such fears. Hopkins (1975) indicated in Senegal that land is rarely rented for more than a year at a time, thereby encouraging its exploitation.

Ecological instability. The ecological stability of ring cultivation is increasingly in doubt. The rising population densities are resulting in an increase in the proportion of permanently cultivated fields, with the remaining fields being left fallow for progressively shorter periods (Marchal 1977). The traditional symbiotic relationship between livestock herders and sedentary crop cultivators is beginning to break down in the face of progressive decreases in grazing land. There is also the question of whether such a relationship can provide the increasing amounts of organic fertilizer required to maintain soil fertility. It has been noted that apart from a few exceptional areas, such as that around Kano in northern Nigeria (Mortimore and Wilson 1965), where manure is transported from the city to the surrounding agricultural areas, the decrease in yields has not been forestalled.

Field fragmentation. Rapidly increasing population densities, combined with the increasing predominance of nuclear family units, is resulting in an excessive degree of field fragmentation.[18] Goddard (1972) found that in the more densely populated area around Sokoto in northern Nigeria, where excessive fragmentation is becoming a problem, farmers operating large farms are spontaneously consolidating fields through various land transactions such as exchange, sale, or purchase. As viewed by most development agencies, fragmentation of fields inhibits the development of agriculture. Certainly that is true with certain types of improved technology. Land improvement and conservation measures may be more difficult because of the need for cooperation among neighbors, whereas small fields may prevent the introduction of mechanization. That in turn has lead to several attempts, through various programs, to encourage consolidation of fields in the region. Jones (1976) documents an example of such an attempt in one area of Mali, an attempt that failed because there was too little sensitivity for tailoring the program to the local situation. On the other hand, there have been occasional successes, even where the impetus has come from outside, of programs characterized by a sensitivity to a local situation and a realization that the local populace has to participate in formulating the consolidation process. Faye and Niang (1977) document such a program in the Experimental Units in

Senegal; another program which has achieved some degree of success is near Fana in Mali Sud.

Combating Increasing Population Densities: Farmers' Strategies.

Currently, farming families are using various strategies to respond to the rising population densities and the decreasing soil fertility levels. Two important ones are using manure and engaging in off-farm employment.

Farming families are becoming more aware of the value of manure for maintaining soil fertility. Increasing tensions between herders and cultivators, concomitant with rising population densities (Baier and King 1974; Horowitz 1972; Bernus 1974; Raay 1975; Diarra 1975),[19] however, are occurring at the same time that the need for manure to maintain soil fertility is increasing. Obviously, the impact of rising population densities depends on the carrying capacity of the land and the seriousness of the problem. The most critical situations are developing in the drier parts of the savanna, where the ecology is particularly fragile and the carrying capacity of the land is low. It is also in these locations where the conflict between herders and cultivators is most intense. That conflict has not been helped by the drought conditions that prevailed in the early 1970s (Campbell 1977). It is one of the paradoxes of the ever-decreasing land/labor ratios that the increasing conflict over whether land should be devoted to crop or animal production inhibits the increasing benefits that livestock can have in slowing the decline in soil fertility. Nevertheless, there is evidence that higher rates of manure are applied as population densities rise. In northern Nigeria, even greater increases were found when the farming families themselves owned the cattle (Table 6.4), even though the cattle owners had bigger farms than did those not owning cattle. Thus, cattle ownership provided the means for greater control over the manure input. Evidence that manure is becoming more of an economic good is found in Gambia and Upper Volta. Transhument cattle owners there are increasingly unwilling to corral their cattle on the fields of other cultivators because they believe they will not be remunerated sufficiently to offset the value they can receive by corralling the animals on fields that they themselves farm.[20]

Off-farm employment, to take the pressure off the land resource, is an alternative way that farming families increase the level of their livelihood. As we have noted elsewhere, the location of the village, the age and gender of the job seeker, and such factors as the capital and skill requirements, determine the opportunities for off-farm employment. Seasonal migration already has been mentioned as a means for widening the scope of such opportunities. In the drier parts of the savanna, in particular, permanent migration has been perceived as a way to overcome the near subsistence level of livelihood obtained from agriculture.[21] Increasingly, however, permanent migration is becoming more difficult because of national political boundaries. Friction between ethnic groups in the region has prevented a more even population distribution. Consequently, some areas are very densely populated; others, thinly populated. Despite those restrictions,

TABLE 6.4
Manure Application, Sokoto and Bauchi, 1966-68

Characteristic	Sokoto[a]		Bauchi[b]	
Population density	High		Low	
Percent owning cattle	38		19	
Cattle owners	Yes	No	Yes	No
Average number of cattle	3	--	10	--
Land (ha's):				
Cultivated	4.1	3.3	3.2	2.7
Fallow	0.1	0.1	2.8	1.3
Per resident	0.8	0.6	0.5	0.5
Man-hours per cultivated ha.	539	679	570	524
Organic manure applied (tonnes):				
Total	16.3	11.4	4.6	0.6
Per hectare	4.0	3.4	1.4	0.2

[a]Average of three villages, in all of which some families owned cattle.
[b]Average of the two villages in which some families owned cattle.

migration has been considerable from certain areas. It has been suggested, for example, that as many as two million Voltaics were living outside Upper Volta in the early seventies--mostly in the Ivory Coast (Kohler 1972; Songre 1973). To what extent does the home community benefit from permanent migration of some of its members? The problem of increased dependency ratios in the home community is often mentioned (Ancey 1974). But that in theory could be offset by remittances from the migrant to the home family. How significant those remittances are is unclear, but in general they undoubtedly are rather low,[22] unless the migrant has a well-paying skilled job in an urban area (Haswell 1975; ORSTOM 1975).[23]

In francophone countries, development workers have viewed animal traction not only as a means of increasing labor productivity but also as a way of maintaining soil fertility. Incorporating residues through deep plowing is the cornerstone for this land-intensification policy (Wilde 1967). Recommended frequencies for deep plowing range from two to four years depending on the soil condition and the rotation being used by the farmer. Charreau (1978) has noted several advantages of deep plowing: it improves water infiltration and soil porosity--which might otherwise limit plant growth; it encourages the conservation of soil moisture not used by the crop; and most importantly, it enables organic matter from harvest residues or fallow to be incorporated into the soil, thereby creating good conditions for its decomposition. Organic matter contributes to improved soil fertility, as witnessed by the increased yields of the succeeding crops of millet, maize, cotton, and to a lesser extent groundnuts. Deep plowing is recommended at the end of the short rainy season, which necessitates planting as soon as the next rains start. Problems arise due to the short period between the end of harvest and the time at which the ground becomes too hard for plowing. In addition plowing is a time-consuming operation. Faye (1978) has indicated that the operation requires four people to line up the residues, another person to work the plow, and yet another individual to lead the oxen. Because such a team takes three days to plow a hectare, which equates with eighteen work days per hectare, it is not surprising that this intensification policy has not been successful to date (Hopkins 1975). Therefore, farmers have tended to see the use of draft animals more as a means of extensification rather than intensification (Milleville 1978; Hopkins 1974; Czarnocki 1973). Substantial empirical evidence verifies this observation. For example, Dunsmore et al. (1976) and Peacock (1967) observed that in Gambia males responsible for growing groundnuts increased the area of groundnuts cultivated per male adult by 18 to 40 percent after obtaining oxen. As population density increases, the pressure for intensification is likely to become greater. That being so, other possible solutions are being tried in the Experimental Units in Senegal (Faye 1977). One possibility is to combine deep plowing and an early harvested crop, such as maize, which not only is harvested relatively early but also gives high yields. Growing crops with these characteristics permits more time for the plowing operation and also may provide sufficient revenue to be used for contract plowing with tractors (Faye 1978). The tractors currently being used, however, can handle only two hectares per day. Two other possibilities are being tried: using larger tractors which, although very expensive, reduce the time constraint and make it possible to plow longer into the dry season; and using tractor-drawn choppers to chop the plant residues, thereby cutting down the manpower from that required for plowing with oxen.[24] The problem of both these strategies is the necessity of using tractors, and one can question the feasibility of doing that extensively in the region at present. Private ownership is constrained by the amounts of capital and land available to individual farming families. Also the history of government-run tractor-hire units in the region indicates that prospects are not promising (Weber 1971; Kolawole 1974).

DISTRIBUTIONAL ISSUES

With the current developmental emphasis on economic growth with equity, a pertinent issue in the West African savanna is what is happening over time to the distribution of welfare within village communities?[25] To examine this complex issue, we look first at changes in villages that create the potential for changes in the distribution in welfare and then at the distribution and composition of incomes. In a final section we examine the potential for exploitive relationships developing among farming families and the possible significance of differences in technical efficiency.

Changes at the Village Level

Underlying much of the analysis in this chapter so far has been the trend toward increasing individualization and the related replacement of a community-oriented social interdependence by an

individually-oriented economic independence, combined with rapidly
increasing population densities. These changes create the
possibility that resources required for producing a livelihood could
become unequally distributed, in that without so much fear of
societal disapproval as formerly, individual gain can be given freer
rein.

Evidence, admittedly of a fragmentary nature, indicates that
partly because of the changes taking place, land is becoming more
unequally distributed, both quantitatively and qualitatively. It
has been observed that increased population pressures tend to
encourage influential groups in villages to increase their relative
share of the cultivated land (Swanson 1978; Dubois 1975), but that
is not always so, as we pointed out in the last chapter. Lahuec
(1970) concluded that in Upper Volta control over irrigated land was
becoming a causal factor in determining increased socio-economic
differentiation at the village level. In addition, distributional
changes can also develop rapidly as a result of differentiation
among farming families in accessibility to support systems and/or
resources needed to purchase capital equipment and improved
technologies. We indicated earlier, for example, that in Gambia, as
concluded by Weil (1970), the introduction of oxen (a new
technology) resulted in landlords demanding the return of land they
previously had rented to other families. It appears, therefore,
that the potential for land becoming more unequally distributed is
increased when improved technology requires a resource, such as
lowland, available in very limited quantities,[26] or has the
potential for encouraging economies of scale, such as oxen provide.

Composition and Distribution of Income

Obviously, if resources are becoming more unequally
distributed, it is likely that incomes are, also. Unfortunately,
apart from Matlon's (1977) research in northern Nigeria, little
rigorous empirical work has been done on the distribution of income
in rural settings. Values derived from Matlon's study (Table 6.5)
indicated that in the southern part of Kano Province in 1974-75
incomes derived from farming were less variable than were off-farm
sources of income. Analyzing incomes from all sources, however,
revealed greater equality in income distribution.

Although in absolute terms the incomes derived from farming
activities were much lower for poor families than for wealthier
families, farm incomes as a proportion of total incomes were higher
for the poor. At the same time, poor farming families earned a
higher proportion of their income as farm laborers for other
families, than did higher-income farmers, presumably because the
poor needed sources of income to overcome seasonal cash and food
problems, and hired farm work was available during that time, and
their needs for additional income were greatest. With respect to
composition of non-agricultural employment, Matlon (1977) found that
the high-income farmers worked at occupations for which remuneration
was higher than average (Table 6.6). Low-income occupations,
generally providing services, required little or no working capital
whereas the number of occupations requiring substantial working
capital increased directly with income category. Of particular

TABLE 6.5
Income Variation Within a Village Setting, Kano, 1974-75[a]

Variable Specification		Income Category[b]					Gini Coefficient
		Lowest Decile	Second Decile	Middle Quintile	Ninth Decile	Highest Decile	
Cultivated area (ha's) per:	Household	2.2	2.4	2.4	2.7	3.2	
	Resident	0.24	0.29	0.41	0.60	0.51	
Average income (N) per:	Family	177.73	234.21	316.57	394.01	626.59	0.3146
	Resident	19.12	28.22	54.27	87.56	99.46	0.2828
Breakdown of income per resident:							
Own farm		80	75	77	60	64	0.3183
Off-farm: Hired farm laborer		8	4	5	4	1	0.5306
Non-agricultural		12	21	18	36	35	0.6097
Cash:							
% of household income generated in or converted into cash		60	51	50	53	55	
% of net cash earned by source:							
Own farm		63	48	53	30	35	
Off-farm: Hired farm laborer		14	8	8	3	2	
Non-agricultural		23	44	39	67	63	
N per consumer:[c]							
Harvest value of:	All crops	31.75	41.25	72.50	106.25	137.50	
	Food crops	13.75	25.00	46.25	60.00	71.25	
Retained food crops[d]		8.75	17.50	35.00	46.25	53.75	
Available food		33.75	41.25	53.75	81.25	90.00	

[a] Matlon (1977) undertook the study in three villages in the southern part of Kano State.
[b] Distribution of income per consumer man-equivalent for each family was divided into deciles and quintiles.
[c] Estimated cost to feed a consumer at a rate of 2,954 calories per day was N47.50 per year.
[d] Included retained food plus food purchases.

TABLE 6.6
Off-Farm Occupations Pursued by Different Income Levels, Kano, 1974-75[a]

Income Category	Occupation[b]	Annual Cash Expenditure per Family (₦)	Average Return per Labor Hour[c] (₦)
Only low income	Shoe repair	6.25	n.a.
	Calabash cutting	0.05	0.09
	Total	2.10	0.09
Low income bias	Provisions trading	41.88	0.07
	Tailoring	37.89	0.20
	Selling grass	-	0.06
	Total	23.24	0.14
Intermediate	Hired farm labor	-	n.a.
	Selling firewood	-	0.13
	Donkey transportation	-	0.13
	Cloth trading	141.64	0.27
	Total	35.06	0.15
High income bias	Local crops trading	314.42	0.26
	Livestock trading	44.53	n.a.
	Processed food trading	252.96	n.a.
	Total	101.97	0.19
Only high income	Petrol trading	123.42	n.a.
	Sack trading	54.78	n.a.
	Total	108.47	0.31

[a]This table is derived from data collected and analyzed by Matlon (1979) in three villages in the southern part of Kano State.
[b]Only a sample of occupations are included in the table but the totals refer to all the occupations sampled in each income category.
[c]n.a. equals not available.

significance was the finding that the distribution of earnings from three food related occupations—trading in local crops and livestock, trading in processed foods, and selling roasted meat—was strongly biased in favor of the high-income groups. All of these activities required substantial annual cash outlays. Matlon (1979) concluded that, as a result, the differences in off-farm occupations tended to widen income disparities by providing profitable investment outlets for the surplus income generated by higher-income families in crop production.

Implications

Unequally distributed incomes within village communities, particularly as they become less cohesive, imply several important developments. Here we briefly discuss three of them: disparity in consumption levels, the possible development of exploitive relationships, and possible differentiation in technical efficiency.

Level of consumption. Data from Matlon's (1977) northern Nigeria study (Table 6.5) imply that at least 60 percent of the farming families in the 1974-75 Kano survey did not produce sufficient food or did not retain enough of their farm-produced food to feed themselves. That meant that they had to enter the market place to purchase food to make up the deficit. At least 20 percent of all the farming families, however, had insufficient income from other sources to use for this purpose. Quite likely each year many more farming families in the West African savanna are faced with seasonal deficits in food supply--a problem commonly called "seasonal hunger" or the "hungry gap" (Benneh 1973; Raynaut 1973; Kafando 1972). Seasonal hunger is characterized by food availability being at its lowest level at a time when the demands of the agricultural cycle are highest and cash resources are also often at their lowest.

Although it is frequently mentioned in the literature as constituting a major problem for farming families in the savanna, seasonal hunger has not often been verified empirically, causing some to question its significance. For example, in the Zaria villages we studied we could not find evidence of its presence, although detailed quantitative data were collected in the consumption study (Simmons 1981).[27] However, we hesitate to dismiss it as a problem in parts of the savanna where the levels of livelihood are lower and/or incomes are more unequally distributed. We simply suggest that further empirical studies are required before categorical statements can be made about its significance.

There is certainly some evidence that seasonal hunger is a problem in some areas. It is obvious that the severity of the period of low food availability during the hungry gap is inversely dependent upon the supplies of food remaining from the previous harvest, the ability to purchase food during this period, and the success of early maturing crops. Let us look briefly at each of these:

1. Results obtained by Haswell (1975) in one village in Gambia in July, 1974, indicated that only 20 percent of the families had enough food grains in store from the previous year's harvest to feed them until the next harvest. Verneuil (1978) mentioned that in precolonial times, granaries held up to a four year supply of cereals, enough to help families during periods of drought.[28] Since then, for various reasons, there has been a decline in the quantities of cereals stored in many parts of the West African savanna. Thus, families were more vulnerable to drought of the early 1970s (Meillassoux 1974).[29] At the same time, that decrease in stored grain would appear to have increased the vulnerability of the poorer farmers to the effects of the hungry season. Such decreases are particularly regrettable not only because the problem of seasonal hunger is potentially greater, but also because there is increasing evidence that storage losses from

traditional methods are considerably lower than those incurred in warehouses operated by official marketing boards (Guggenheim 1978). Both Guggenheim (1978) and Giles (1965) have found that losses of millet and sorghum in traditional storage systems are not usually more than 4 percent per year.

2. An important traditional way of lessening the effect of the hungry season has been to grow early maturing crops such as hungry rice (Digitaria exilis) and early millet. Changes, however, have taken place in some areas where increased emphasis on cash crops has encouraged the demise of these early-maturing types which often have relatively low yields (Norman, Newman, and Ouedraogo 1981). Also changes in responsibilities within the family and in the availability of labor appear to have had an impact in certain areas. For example, it has been suggested that in Gambia the production of early millet has been discouraged by the shift on the part of women to grow swamp rice (Weil 1973) and by the increasing unavailability of children to send to the fields to keep birds from damaging the crop.[30]

3. Purchases of food provide another means of alleviating the effects of the hungry period. Unfortunately, if families are not self-sufficient, food purchases are needed when cash resources are lowest (Dunsmore et al. 1976) and prices highest (Steedman et al. 1976). Although at such times in the agricultural cycle opportunities for hired farm work are greatest, such work and other off-farm work pursued by low income individuals are, as we earlier indicated, relatively low paying. Also such work has to be undertaken by those who need it at a time when demands on their own farms are highest. Under such circumstances, an alternative strategy is to borrow money--a strategy that can have its problems, some of which we will shortly discuss.

The effects of the hungry gap on the physical constitution of the hungry have seldom been examined in detail. One of the few exceptions was Haswell's (1975) study in Gambia in which she, through interdisciplinary work with medical personnel, correlated the agricultural cycle to nutritional levels. Results indicated that during the peak of the agricultural cycle the individual's calorie and protein consumption was lowered (Grant 1950), while loss in weight tended to occur because of the reduced intake in relation to the increased working burden (Platt 1954; Hunter 1966). A potential for further debilitating effects was thus created as the chances of contracting nutritionally related diseases were increased and the body's resistance to other illnesses was decreased (Chambers and Longhurst 1979).

Exploitive relationships. Francophone literature, and increasingly that emanating from anglophone countries, cites the development of exploitive relationships within villages, and between villagers and those from outside the village. In view of the social changes that are occurring, there does appear to be increased potential for exploitation. As logical as that may seem, however, hard empirical data are generally lacking. Therefore, we are forced to present evidence of an indirect nature.

We have already cited some evidence that, under certain circumstances with increasing population pressures, land does become more unequally distributed. That combined with differential access to support systems and thus to improved technologies, increases the likelihood of widening income disparities. Hence, the potential for exploitive relationships is increased between those who have resources for production and those who do not. Two oft-cited examples of such relationships developing are as follows:

1. Dependency relationships develop between farming families who own draft animals and those who do not, when services provided in the form of plowing are paid for by labor. In Mali, for example, Ernst (1976) found three to five days of labor were expected in return for one day's plowing. Jones (1976) cites work done by Gallais in the interior Niger delta which revealed that land preparation was not a major bottleneck. Even so when the plow was introduced, individual families not owning plows requested neighbors who owned them to plow their land.[31] They paid for the plowing by giving manual labor during the bottleneck periods of weeding and harvesting, so they were forced to cultivate less land themselves. The plow owners, by being able to obtain labor at the bottleneck periods by plowing the fields of others during the period when labor was not so limiting, were able to increase the areas they cultivated. That, of course, could result in widening income disparities.

2. The lack of food self-sufficiency and the problem of seasonal hunger also have the potential for creating dependency relationships. One way reportedly often used to obtain food, but which was not confirmed in the Zaria villages studies, was to borrow money or food for consumption during critical periods such as the growing season. Hierarchical trading systems facilitate doing that, thereby providing a means for exploitive relationships to develop (Raynaut 1976; Verneuil 1978). Chambers and Longhurst (1979) have pointed out the contribution of this hypothesized "seasonal screw" to the low-income poverty trap (Matlon 1977). The impoverishment of certain groups in a society can become even more severe when means of production other than labor are used to buy survival. For example, in severe times agricultural materials purchased on credit can be sold at very low prices (Venema 1978). Another example would be the pledging of land. Such strategies can lead to a "ratchet effect" (Chambers and Longhurst 1979) with a downward spiral. In other words, survival commitments one year lead to a lower potential income and a higher level of indebtedness for the next year. This ratchet effect can be reinforced by shock events such as drought (Charlick 1974) or death in the family (Chambers and Longhurst 1979). Therefore, it has been suggested that lack of food can

provide the milieu in which poverty is sustained and deepened.

Technical efficiency. It now appears that recent preoccupation with allocative efficiency[32] has detracted from the importance of looking also at technical efficiency[33] as a determinant of economic efficiency. In an analysis of differences in allocative and technical efficiency among farmers of northern Nigeria, Matlon (1981) found that poorer households were disproportionately represented among the least technically efficient producers.[34] Technical efficiency was found to decline rapidly if the planting of sorghum and late millet, as well as the first weeding, were delayed. The degree of intercropping, as reflected by a greater number of crops per mixture, was found to be positively associated with greater technical efficiency. Matlon (1981) has pointed out that identifying differences in technical efficiency is neither a necessary nor a sufficient condition for demonstrating interfirm differences in managerial ability. Such an inference is valid only if farm managers share common objectives and face the same range of production choices and also the same external constraints. The latter include both the technical element and the exogenous factors of the human element. The history of the farming family--the status, income, and liquidity position that family has inherited from a previous period--determine access to resources and thus influence the production and employment strategy the family adopts.

Analyzing the management practices by families in different income groups revealed that the degree of intercropping was the single management factor in which poor households exhibited more technically efficient behavior than did the wealthier. For the crops examined--millet, sorghum, and cowpeas--poorer families tended to plant somewhat later than average; rather than indicating a lack of managerial competence, however, that later planting might have been deliberate by families who conceivably were short of both cash and seed. Farmers who plant early in the rainy season risk the necessity of replanting if the first heavy rains are followed by too long a dry season. The greatest difference in technique between the low- and high-income families was found to be in weeding. Low-income farming families first weeded their fields six weeks after the start of the rains--almost two weeks later than high-income farmers. Second and third weedings by low-income farming families were about half of those undertaken by the families in higher income groups. Once again, it is tempting to attribute those differences to management and/or motivational differences. An alternative explanation might be that the urgent need for food and/or cash made it necessary for members of low-income families to work at off-farm occupations to provide the means for sustenance and to repay debts.

IMPLICATIONS FOR CONSTRUCTIVE CHANGE

Different changes over time, as well as current differences in factors associated with the technical and human elements, help to account for the diversity that exists in farming systems across the West African savanna. Some of these factors also contribute to the increasing differences in household welfare which appear to be emerging within many village communities. Design of relevant agricultural policies, support systems, and improved technologies is a crucial ingredient in rechanneling those changes constructively for the benefit of the whole society. By the same token, poorly designed agricultural policies, support systems, and improved technologies could adversely affect the society as a whole, particularly for certain parts of the farming community, by exacerbating income disparities within villages and communities. As we have frequently indicated, we believe that the farming systems approach to research is potentially a powerful analytical tool for helping to develop and test relevant agricultural policies, support systems, and improved technologies.

In conclusion, we specifically mention two issues that must be solved in order to determine which direction the development of improved technology should take in the West African savanna:

1. Should research emphasize developing improved technologies for mixed cropping? We believe it should, particularly for food crops, despite the demise of mixed cropping in some areas. We base our view on increasing empirical evidence which demonstrates the potential for growing crops in mixtures by using improved technology (Baker and Yusuf 1976; Kassam 1973; Baker 1975 and 1979; Kowal and Kassam 1978). Most of the work to date demonstrates that such systems can make better use of the technical environment, a crucial benefit in the resource-constrained savanna. It is also likely that mixed cropping systems can be better adapted than sole cropping systems to the human environment in the region. At the same time, we believe that the move to sole crops has not been due to the intrinsic superiority of sole crops under improved technological conditions but rather due to pressures exerted by researchers and extension services; technological development has proceeded in conjunction with the idea of growing crops in sole stands.

2. What should be done with respect to the problem of rising population densities and the apparent competitive relationship between crops and livestock? Perhaps first it should be determined what is needed to sustain soil fertility. We cannot adequately answer that. Although there does appear to be a great deal of evidence that animal manure will continue to be an important source, there are varying estimates of the amount of animal manure required to permit permanent cropping. For example, Guinard (1967) has indicated that in the West African savanna areas, 10 tonnes of manure per hectare are required annually to permit permanent cropping of millet or sorghum. On the other hand, Alkali (1969) indicated that in northern Nigeria 2.5 tonnes per hectare per year are sufficient to maintain yields in most areas. Nonetheless, although in the special case of Kano, mentioned earlier, farmers reportedly apply up to 5 tonnes per hectare, it is doubtful that these levels of application generally could be sustained, given the current relationship between herders and cultivators. That prompts us to ask another question: if cattle could be more firmly

198

integrated into the cultivation system would cultivators
themselves be willing to undertake both ownership and
management functions? If yes is the appropriate answer,
that would imply the need for very fundamental changes in
some of the farming systems in the region.[35]

NOTES

1. Elsewhere we have couched this discussion in terms of the
horizontal and vertical dimensions--with the latter being divided
into the historical subdimension in terms of the past and the
prospective subdimension in terms of the future (Norman, Newman, and
Ouedraogo 1981).
2. It may also be a response to the lower fertility of the soil
compared with the Zaria area. But such an observation would be
contrary to that found by Lagemann (1977), who concluded that in the
much wetter area of eastern Nigeria--outside the savanna or
semi-arid region--farmers tend to plant more densely when the soil
fertility declines. The difference between that practice and the
one we suggest could perhaps be explained in terms of the much
higher potential for soil moisture stress in the semi-arid area.
3. Nevertheless, one point that should be emphasized is that
although the grain yields of individual crop constituents might be
depressed, they can still have considerable economic value to the
farmer. This applies particularly to the cowpea haulm, which
provides food for livestock. Estimates of the value of the haulm
were omitted from the analysis of the individual crop enterprises.
4. In the nine Bauchi, Zaria, and Sokoto village samples as a
whole, only 22 percent of the farming families were found to own
cattle (Norman, Pryor, and Gibbs 1979).
5. Other studies have shown that in fact cattle ownership by
crop cultivators tends to be concentrated in the hands of wealthy,
influential families (Dunsmore et al. 1976; Clough 1977).
6. The following countries, those with more than one million
population and with some portion in the savanna, were included in
the calculations: Benin, Cameroon, Chad, Ghana, Mali, Niger,
Nigeria, Senegal, and Upper Volta. Gambia with less than one
million people was not included.
7. Some writers have questioned the lack of exploitation in
some of these traditional societies (Ernst 1976; Kafando 1972).
However, for reasons we discuss, we believe that the potential for
exploitation is likely to be much greater in the future.
8. To date the spread of Islam throughout the savanna has
probably mitigated against the inequalities developing as rapidly as
they could.
9. Power may also be abused by new power groups in the
villages, such as traders and money lenders, who hold economic power
over the more disadvantaged groups without the responsibility
embodied in the patron/client relationship traditionally
characteristic of the communities (Haswell 1975; Murphy and Sprey
1980).
10. It is perhaps significant that in a similar ecological area
in India, Binswanger, Jodha, and Barah (1980) found little

difference in the attitudes of small and large farmers operating small and large farms. They therefore concluded there was no need to develop risk-graded improved technologies. Rather the answer to the risk problem lay in developing more equitable access to external support systems.
  11. These were French owned in the earlier days but increasingly they are parastatal organizations under the jurisdiction of the independent country. For example, in Mali Sud the parastatal CMDT has evolved from the French cotton organization CFDT.
  12. It is therefore not surprising that a recent analysis undertaken by the World Bank (1981a) indicates that in the savanna areas of West Africa export cash crops are currently more efficient at converting domestic resources into foreign exchange than food crops are in saving foreign exchange. However, changes in price relationships and development of improved technologies for food crops could rapidly change the situation.
  13. This could well imply that animal traction, in the form it is currently used in West Africa, is incompatible with mixed cropping. Animal traction does reduce the possible combination of crops that can be grown. For example, with animal traction it is not possible to grow crops on the ridge and in the furrow, as is sometimes done by using the hand-labor systems in northern Nigeria. However, animal traction does not appear to eliminate the potential for mixed cropping under practical farming conditions (Unité d'Evaluation 1978). There is also no evidence to suggest that mixed cropping is incompatible with mechanization involving animals. Andrews (1972) has demonstrated that slight modifications to cropping patterns can correct an apparent incompatibility. However, it is likely that over time the impact of animal traction has negatively influenced the relative dominance of crop mixtures in the region.
  14. This problem, sometimes called the "hungry season" or soudure in French, is discussed later in the chapter.
  15. In an attempt to keep prices low for people in urban areas, prices given to producers in many francophone countries for food crops marketed through the official system have been well below those received in the parallel (unofficial) market (Harriss 1978). Not surprisingly many producers have produced food crops mainly for fulfilling household consumption needs.
  16. Therefore, analysis of the labor-flow data in aggregation periods of less than a month would probably accentuate the labor bottleneck period for weeding relative to that of harvesting.
  17. This is often the first equipment supplied when oxen are adopted by farming families.
  18. It has also been suggested that fragmentation is accelerated in areas where the Maliki rather than the customary law dominates (Dunsmore et al. 1976).
  19. This is due to decreases in grazing areas not only on rainfed land but also on lowland where cattle traditionally have grazed during part of the dry season. Such land is now, of course, increasingly being converted into irrigation schemes for growing crops--for example, Mali, Senegal, Gambia, and Upper Volta.
  20. It should be noted that herdsmen on occasion do cultivate

their own fields, particularly in areas that are densely populated.
21. We recognize that factors other than declining soil fertility levels could contribute to this. In addition, factors such as the attractiveness of cities and settlement schemes may encourage permanent migration.
22. Amin (Campbell 1977) has stressed the loss of potential national income that occurs as a result of individuals migrating from the land-locked Sahelian countries to countries having coastal areas.
23. Caldwell (1968) presents a viewpoint at variance with this by arguing that rural-urban migration in Ghana has raised rural living standards.
24. Alternative possibilities with such chopped material would be to spread it over the surface, make it into compost, or feed it to animals, which would result in animal manure.
25. We recognize that equally important is the distribution of welfare among different rural areas and different sectors, that is rural and urban. However, discussion of the former has been covered at least implicitly, earlier in the chapter, while the latter impinges on policy issues of overall economic development, many of which are beyond the scope of this book.
26. Earlier, we mentioned the example of village leaders taking control of the limited amounts of lowland so as to reap the benefits from the improved technological package for dry season tomato production (Agbonifo 1974).
27. Also Rowland et al. (1981) working in Gambia found that simply giving food grain to farming families was not sufficient to solve the problem. Rather it had to be given in cooked form, indicating that women did not have time to cook because of agricultural activities!
28. Also Guggenheim (1978) has documented the decrease in the number of traditional granaries in Mali.
29. Apeldoorn (1981), in analysing the impact of the drought in northern Nigeria, concluded that the rural majority has been made less able to deal with such crises because economic developments since the beginning of the colonial era have taken away the automony and intrinsic coherence of the traditional structures. The self-help strategies which saved them in the past have been replaced by widespread dependence on government.
30. This is because of the increasing frequency of secular education for children.
31. The reason for that is not entirely clear, although difficulty of preparing the land by hand, or the critical nature of timing of the operation, might be important. If so, however, land preparation would be a critical bottleneck for those not owning plows, which would make their strategy more rational.
32. Encouraged by the landmark work of Schultz (1964).
33. Technical-efficiency differentials are the variation in output across a set of firms using the same combination of inputs not caused by differences in technology or by random disturbances.
34. Mijindadi (1980) and Pendleton (1980) also have examined these issues by using data from the various village studies. See also Matlon and Newman (1978).
35. Delgado (1978) has investigated this possibility in some

depth but came to the conclusion that under current conditions it was not very feasible.

7
Improved Technology: Assessing Suitability by Using a Farming Systems Approach

> "Without fine tuning new production methods to fit the physical and socio-economic environment, the probability of farmers' adoption will be severely reduced and the benefits derived from investment in agricultural research and extension will only be a fraction of their potential."
>
> H. Zandstra (1979a)

Savanna farming systems, as revealed in the preceding chapters, differ considerably in many dimensions, but in one dimension they are alike--most are relatively unaffected by modern agricultural technologies. There has been little use of improved seeds, inorganic fertilizers, herbicides, pesticides, or mechanical cultivation or harvesting equipment, even though for years the agricultural research system and extension services have been recommending their increased use. And all concerned with Nigeria's development agree that the future growth of the country is tied closely to agricultural development and to the introduction and use of relevant improved technologies. Productivity of labor -- particularly at bottleneck periods--and the productivity of land must be improved to meet the demand for both food and jobs for a rapidly growing population. Furthermore, the productivity of the fragile savanna ecosystem must be increased on a sustainable basis if farming systems are to survive for more than a few years.

Meeting this agricultural challenge in a manner compatible with the needs of individual farming families and also the society as a whole requires that three basic questions be addressed by agricultural researchers:

1. How can improved technologies be efficiently designed and developed to provide relevant, practical solutions to the problems of farming families?
2. How should the improved technologies be packaged and the necessary external institutions or support systems be organized and put in place to ensure that farming families will adopt them?
3. How should the potential conflict between short-run private interest and long-run societal cost be avoided and/or resolved?

We have suggested repeatedly that by adopting a farming systems approach, researchers can respond with increased effectiveness. In this chapter, we draw on experience at the Institute for Agricultural Research (IAR) to discuss how FSR, as applied to two quite different research programs, can help answer the questions. The first program involved the testing of four technological packages on farmers' fields. Although descriptive/diagnostic work presented in Chapter 5 already had been completed when these field tests were begun, the design phase of the technological packages had been undertaken independently of the diagnostic phase. So the testing of the packages was neither so efficiently nor so holistically done as it might have been if a farming systems approach to research had been envisioned at the outset. Nevertheless, the testing partially illustrated the elements of FSR in that: first, it involved the farmers as more than laborers and owners of land resources; second, the team of researchers included both technical and social scientists, so was multidisciplinary; and third, considerably more than yield-per-hectare variables were monitored. As shown by the discussion of results below, the recommendations were modified to reflect the farm-level experience.

The second program involved action research to explore more fully how adoption of new technology packages was influenced by three variables relating to the support systems: first, the assured availability of improved inputs; second, the availability of formal credit; and third, the utility of increased information flows.

Both experiences demonstrated, in our view, that the farming systems approach to research can greatly increase the understanding of the suitability of improved technologies and can suggest possible changes in support systems. Some answers to the first and second of the three basic questions can be suggested on the basis of those experiences, particularly in that exogenous and endogenous factors outweighed many of the anticipated technical constraints to the farmers' likely adoption of the technologies. The third question, unfortunately, was only obliquely addressed by the experiences, although it is a very important issue.

TESTING IMPROVED TECHNOLOGIES

The improved technology packages tested were developed by technical scientists working at IAR. All were oriented primarily to relieving biological constraints to increased yields on rainfed crops. The sorghum and cotton packages were being recommended to farmers in the area at the time of the testing, and the maize and cowpea packages were almost ready to be recommended on the basis of extensive experiment station development and testing.[1] Though none had been formally tested in on-farm situations,[2] all packages were developed to increase the productivity of land. Because it was not the practice at IAR to measure labor use or to calculate marginal returns to labor, output per unit of land could be termed the primary performance criterion for technological development. In the on-farm tests reported here, three sets of evaluation criteria were applied: technical feasibility, compatibility with exogenous factors, and compatibility with endogenous factors.

Package Requirements

The requirements of the various packages tested are summarized in Table 7.1. Some of the major ways they differed from traditional or indigenous practices were as follows:

1. Although suggested planting dates for sorghum and cowpeas were essentially the same as under traditional practices, the date for cotton, a nonfood crop, was in mid-June, two to four weeks earlier than under indigenous conditions. The suggested planting date for maize, a crop not widely grown under traditional systems in the area, was similar to the date for sorghum.
2. All the tested packages involved growing crops in sole stands, which traditionally has rarely been done in the Zaria area,[3] although sole cropping is more commonly practiced by farmers who use animal traction rather than just hand power. In any case, plant-stand densities suggested were much higher than for crops planted in sole stands by using traditional practices.[4] The thinning operation, therefore, was an important component of the suggested improved technologies.
3. Improved seeds, together with seed dressing for sorghum, maize, and cotton, were used for all packages. Because cotton seed was distributed by the marketing board each year,[5] the cotton variety was the same as that grown under traditional practices and in addition normally had already been dressed.
4. Substantial amounts of fertilizer, though seldom used by farmers under practical farming conditions, were suggested. Much fertilizer was recommended for maize, but none was suggested for cowpeas.
5. Spraying for pests was an important component of the suggested packages for cotton and cowpeas.

Setting up the Tests

Table 7.2 gives some details on implementing the tests at two locations: the cowpea package in one of the Zaria study villages, Hanwa, where hand power was used; the other packages (cotton, sorghum, maize) at Daudawa, about 80 km northwest of Zaria, where farmers were using animal traction.[6]

Certain exogenous variables were altered considerably from the normal situation for the tests. In addition to the obviously greater and more direct information flow between researchers and farming households, the North Central (now Kaduna) State government provided extension staff and improved inputs. A credit program for the improved inputs involving group responsibility for repayment was also instituted; in 1974 it involved lending to individuals at 5 percent interest with repayment at harvest. No adjustments were made in the normal marketing arrangements for cotton, sorghum, and cowpeas. But because maize had not been grown traditionally, a guaranteed price was offered to farming households wanting to sell

TABLE 7.1
Suggested and Actual Levels of Improved Inputs Used by Test Farmers, Daudawa and Hanwa, 1973-75[a]

Variable Specification	Units	Cotton[b]			Sorghum		
		Technology			Technology		
		Improved		Indigenous	Improved		Indigenous
		Suggested	Actual	Actual	Suggested	Actual	Actual
Size of plot	ha	1.2	1.3	1.7	0.6	0.5	2.0
Inputs:							
Units of N:P:K	kg/ha	32:23:0	27:22:0	1:0:0	98:45:0	95:46:0	0:0:0
Seed:							
Amount	kg/ha				13.6	13.2	10.3
Variety					SK 5912	SK 5912	Local
Seed dressing:							
Amount	packets/ha				4.9	5.1	1.1
Type					Aldrex T	Aldrex T	Aldrex T
Spraying (nos.):		6	4	1			
Plants:							
Planting date		15 June	21 June	5 July	Start of rains	23 May	26 May
Stands/ha[c]		28.6	23.5	21.3	43.9	23.0	12.9

		Maize		Cowpeas	
Variable Specification	Units	Improved Technology		Improved Technology	
		Suggested	Actual	Suggested	Actual
Size of plot	ha	0.4	0.4	0.2	0.1
Inputs:					
Units of N:P:K	kg/ha	198:50:50	189:49:49	0:0:0	0:0:0
Seed:					
Amount	kg/ha	17.3	17.8		
Variety		S123	S123	Acc 593 (bulk)	Acc 593 (bulk)
Seed dressing:					
Amount	packets/ha	7.41	7.61		
Type		Aldrex T	Aldrex T		
Spraying (nos.):				6	6
Plants:					
Planting date		Start of rains	4 June[d]	25 July[e]	1 Aug[e]
Stands/ha[e]		43.9	19.3[d]	33.3[e]	22.6[e]

[a]See Table 7.2 for years referring to specific crops.
[b]Seed and seed dressing were distributed free by the marketing board so no records were kept of amounts used.
[c]In 1,000 units.
[d]These results were for 1973 only.
[e]These results were for 1975 only.

TABLE 7.2
The Technology Packages and Weather During the Years in Which They Were Tested, Daudawa and Hanwa, 1973-1975

Variable Specification	Daudawa (Oxen Power) 11°38'N 7°09'E			Hanwa (Hand Power) 11°8'N 7°43'E		
	1973	1974	Long Term Average	1974	1975	Long Term Average
Numbers of farmers who tested the technological package[a]:						
Cotton	19	23		–	–	
Sorghum	19	24		–	–	
Maize	19	20		–	–	
Cowpeas	–	–		10	10	
Weather:						
Rainfall (mms.)	594	1176	1082	1115	988	1115
Months with surplus rainfall	Aug.-Sept.	July-Sept.	June-Sept.	June-Sept.	June-Sept.	June-Sept.
Growing season[b]:						
Length (days)	153	185	174	200	190	180
Start	May 21-30	May 1-10	May 11-20	May 1-10	Apr. 21-30	May 11-20
End	Oct. 21-30	Nov. 1-10	Nov. 1-10	Nov. 11-20	Oct. 21-30	Nov. 1-10

[a] Packages not reported in this table were: one for maize that required less fertilizer and was tested in Daudawa (Norman et al. 1976a); one for cowpea undertaken in Doka in 1976; and one for cotton tested in 1971 and 1972 in two other villages where only hand power was used (Norman, Hayward, and Hallam 1974 and 1975). The cotton package was also tested in 1971 and 1972 in Daudawa, the results of which are averaged in, in calculating Table 7.4.
[b] See the footnote to Table 3.2 for definitions of "length", "start" and "end of growing season."

some of their corn crop.[7]
Because of the way in which farming families were selected, possibly a biased sample resulted. Farmers included were those who, after attending a village meeting on the proposed project, indicated an interest in participating in the testing program. Whereas it could have been expected that the participating farmers all would have been uniformly progressive or would have had above average resource endowments, that did not in fact appear to be the case. For example, in Hanwa the farmers participating in testing the cowpea package did not appear to differ significantly, in terms of many variables, from the average Hanwa farmer in the farm-management study discussed in Chapter 5.[8] There were, however, significant differences in the attitudes and performances among the farming families participating in the testing program. Indeed, quite likely some farmers expressed interest in the project because it ensured their access to improved inputs, which were generally in short supply, rather than because they were genuinely interested in considering the potential of particular improved technologies. Although no crop insurance was involved, those chosen as test farmers achieved a certain status. Great effort was made to extend the opportunity to participate to a wide range of farming families. That was particularly important because, to get an idea of the robustness or potential distribution of the improved technology, it was necessary that many farming families participate under a wide range of conditions in these farmer-managed tests.

The initial pressures for undertaking farm-level testing of technological packages came from technical scientists within IAR. The scientists working on cotton were particularly concerned about the general lack of adoption of the cotton technology package being extended to farming families, and they wanted to identify possible problems with the package, so adjustments could be made to encourage better adoption. Scientists working on maize were particularly interested in farm-level testing of a promising maize package; those involved with cowpeas were anxious to examine their proposed cowpea package under farm conditions in which the pest complex might be substantially different from that at the experiment station. The sorghum package was an obvious technology to investigate because of the significance of the crop in the economy of northern Nigeria. The teams involved in the farm-level testing consisted of both technical and social scientists, who worked well together in an interdisciplinary mode.

The participating farming families were asked to try the improved technologies on plots of a specific size (see Table 7.1). Agricultural assistants, who were the extension agents, encouraged, but did not force farmers to follow the suggestions for the improved technologies. Enumerators from IAR recorded details of daily activities--inputs and outputs--not only on the plots where the improved technologies were tested but also on fields on which the test farmers were growing the crops by using indigenous or traditional practices.

Evaluating the Packages

Technical feasibility and compatibility with exogenous factors were assumed to be the necessary conditions for the evaluation process; that is, if the technologies were not technically feasible or could not be adopted within the framework of exogenous constraints--such as prices or transport--they were judged to be unsuitable for farmers' conditions. Compatibility with endogenous factors was assumed to be the sufficient condition; that is, though the packages might be technically suitable and reasonably well matched with exogenous variables, assessing the endogenous constraints would enable predictions to be made on how large the group of potential adopters might actually be. Using both exogenous and endogenous compatibility as criteria obviously reflected a short-term perspective. Over time, changes in exogenous factors (such as the availability of credit or transport) could affect significantly the assessment of compatibility. To permit a long-term perspective, institutionalizing the farming systems approach must be considered; we discuss that topic later.

Technical feasibility. Testing for the different improved-technology packages extended for various periods between 1971 and 1976 and included the drought conditions of the early 1970s. Capturing variability in rainfall was important because it enabled yield stability or risk to be examined under rainfed farming conditions. Only half the normal rain fell in 1973, and the growing season itself was considerably shortened (Table 7.2).

The improved maize, S123 composite, with a growing season of 120 days, fit well within the growing season even during the unusually dry year of 1973. It thus met the basic requirement of technical suitability. Each year the yields for the improved maize were similar, with coefficients of variation of about 40 percent (Table 7.3).

Average yields of the improved sorghum variety, SK5912, with a 160-day growing season, were considerably more than double those of indigenous varieties, but variability of yields was also greater with the improved sorghum. The average yields of the traditional and improved sorghum in 1974, a year of relatively favorable rainfall, was 73 percent higher than in 1973, a dry year.[9] Recognizing that one way to reduce the variability in total production due to variation in the length of the growing season would be to develop a shorter-season sorghum, scientists at IAR have been involved in substantial research to develop such varieties; to date, however, those developed have been susceptible to head mold (see Chapter 3). To reduce the importance of the variability of sorghum yields--although not the variability itself--farmers plant the sorghum in mixtures with millet. In that the yields of the short-season gero millet generally are not affected by water stress, the variability in total grain output per hectare under rainfed conditions is reduced by the mixed planting. Scientists at IAR have now followed the farmers' lead and are conducting experiments with millet/sorghum mixtures; experimentation is being officially extended under the auspices of the National Accelerated Food Production Program.

Still, technical evaluation of the improved maize and improved sole-crop sorghum packages showed that, based on yield factors alone, farmers would prefer to grow the improved maize, the yields

TABLE 7.3
Variability in Returns from Improved Technology Packages, Daudawa and Hanwa, 1973-75

Power Source	Crop	Days Between Planting and End of Growing Season	Yield Average (kg/ha)	Yield Coefficient of Variation	Percent of farmers who Covered Costs[a]	Net Return More than Average for Indigenous Practices	Net Return More than Average for Improved Cotton Package	Net Return (N/ha)
Oxen	Cotton:							
	Indigenous:							
	1973	91	454	32	88	50	50	16.72
	1974	131	364	55	100	42	17	38.84
	Improved:							
	1973	110	658	40	79	42	37	16.60
	1974	143	734	37	100	78	48	80.13
	Sorghum:							
	Indigenous:							
	1973	128	488	55	83	42	75	37.95
	1974	185	345	40	89	44	22	52.07
	Improved:							
	1973	141	1161	69	100	53	63	80.77
	1974	179	1530	38	100	67	42	82.46
	Maize:							
	Improved:							
	1973	129	2867	37	100	--	100	193.96
	1974	167	2927	43	100	--	75	186.75
Hand	Cowpea:							
	Improved:							
	1974	111	1534	40	100	--	90	199.00
	1975	94	453	37	0	--	0	-77.42

[a] Included value for family labor, and where applicable, subsidized costs for the improved inputs.

of which were not only higher but also generally more dependable. Cowpeas showed considerable differences in average yields between years (Table 7.3). Yield variability, however, was attributable to the increased presence of coreid bugs (<u>Acanthomia brevirostris</u>) during the second year of field trials, rather than to the impact of rainfall variations. Applying DDT/BHC insecticide in a water-based form the first year and in an oil-based formulation the second year did not effectively control the coreid bugs, which had not emerged as a major pest on earlier experiment station-based work. Though the spraying recommendation was modified midway through the second year, by using endosulphan for the last two or three sprays (in a six-cycle spray), the change occurred too late to save the crop. When a third year's attempt with the cowpea package again resulted in pest-control problems, the package was referred back to technical scientists for further on-station research.

The recommended earlier planting date and the use of sprayers for pesticide application were, as we indicated earlier, major changes involved in adopting the cotton package. For the most part, the participating farming families did try, though somewhat reluctantly, to plant the test plots near the suggested date, but it was apparent that their food-crop activities had much greater priority.[10] The long-term solution in such a situation might be to develop a variety and a set of recommended practices that would permit later planting. The problems of water--a lack of nearby water sources and thus difficulty in transporting water to cotton fields for use in the spraying operation--were easier to solve. The magnitude of the potential problem was indicated by the fact that the spraying operation had to be undertaken weekly for six weeks, beginning nine or ten weeks after planting, by using 225 liters of water per hectare per spraying.[11] In the third year of testing, a switch was made to an oil-based insecticide and an ultra-low volume (ULV) sprayer. The degree of pest control was the same, provided that all six sprayings were undertaken. Eliminating the need for water, the collection of which constituted 26 percent of the time needed for spraying with water-based insecticide, combined with a 23 percent reduction in actual spraying time made the ULV machine especially attractive. In addition its relative cheapness, light weight, and ease of operation made the alternative spraying system much more feasible to adopt (Beeden, Hayward, and Norman 1976).

<u>Compatibility with exogenous factors</u>. Among the exogenous factors taken into account in the on-farm testing were a range of acceptability, marketing, processing, and labor conditions. As already mentioned, the availability of information and inputs had been altered for the tests and thus could be considered only by making comparisons with conditions of less well-served villages.

The improved sorghum and cowpea varieties were not visibly much different from local varieties, so there were no market or taste constraints. Similarly, the cotton variety was readily marketed through normal channels. The S123 maize composite, which had hard kernels and hulls that were difficult to remove, was not so easily accepted for household consumption or for marketing, however. Hand grinding was difficult; village engine operators charged more for grinding it than for grinding other grains mechanically. Different varieties, different methods of food preparation, or stronger grinding machines could, over time, reduce those constraints. Although maize's yield and profitability were potentially very high, neither the local food market nor the feed grain markets were well developed.[12] Thus, to increase the suitability of the maize package in the short run, some overt participation of the government in maize marketing would be necessary.

With the cowpea tests, there were some exogenous constraints in the form of farmers' risk assessments and labor supplies. The test, for example, required sole cropping in direct contradiction of most farmers' firm beliefs that cowpeas will not do well planted alone. Without spraying, of course, their concern was scientifically justified (Raheja 1976). Even though the test specifically involved spraying, farmers who had agreed to set aside 0.20 ha for the test planted only 0.12 ha. Another constraint was the high input of labor required for harvesting; cooperating farmers accepted that with difficulty. Fifty-six percent of the total labor input was for harvesting, particularly in November when labor demands for picking cotton were also peaking. Women and children provide the major source of labor for harvesting both cowpeas and cotton: therefore, if the area devoted to either crop or both were to increase, pressure on the institution of wife-seclusion would increase.

Questions on adequacy of input and on information-supply systems were raised indirectly, if not measured directly, by the tests. As improved technological packages become more significant components of the farming system, it is obvious that the significance of timely delivery of the right inputs will increase. The input distribution system prevailing at the time of the tests simply could not provide the substantial quantities of improved inputs, particularly fertilizer, the maize package required. In addition, the cotton and cowpea packages depended on adequate spraying materials, and all packages required new information. At the time of the testing, extension concentration in that area of Nigeria was clearly insufficient: about one extension agent per 2,500-3,000 farmers. Except for maize, adopting the improved technologies would involve a drastic change from mixed to sole cropping. As revealed by the testing, some operations were sensitive to timing, and spraying cotton and cowpeas involved a relatively complex technology. Also management improvements might have been needed; at least the significance of good management was reinforced by the need to minimize the risk of a low payoff attached to the high investment in money and hours of labor.

Finally, there was the exogenous constraint of credit. In addition to problems of physical access to inputs, before adopting the improved technologies, farmers would have to solve the problem of obtaining the substantial cash required for their purchase. An efficient credit program undoubtedly would facilitate adoption. In our tests individual farming families received loans for improved inputs on the test plots, and a repayment program for the in-kind credit involving some group-responsibility for loans proved to be successful.[13] After some discussion at the outset of the project, it was agreed that a 10 percent surcharge would be levied on all credit, against possible default. Because repayment was virtually complete in both 1973 and 1974, substantial refunds were made in both years.

Thus, an important necessary condition for the adoption of the packages by farming families was the presence of a strong support system. In fact, in the Daudawa area shortly after field testing was completed, a stronger support system was initiated through the Funtua Agricultural Development Project.

Compatibility with endogenous factors. Ultimately, technology packages of the types tested in Nigeria must be fitted into a farming system and must be judged in terms of meeting that farming system's goals. As we have described earlier, Zaria farming households appeared to be labor-short and risk-averse, but open to opportunities for profit maximization subject to a food-security constraint. That set of constraints and objectives reflected the need to guarantee food supplies, to maximize returns to labor, and to recover cash involved in undertaking any technological changes. It also suggested that, particularly for farming households with resources barely adequate for current food needs, any technological changes made would be incremental. Thus, those improved technologies requiring only one or two changes in the system would be more likely to be adopted than those requiring a whole series of modifications.

With the recommended cotton variety, for example, farmers would be required to plant in June instead of July. That implied that food crop planting schedules would have to be altered significantly to accommodate the early-planted cotton. Analysis of the test data, however, showed that farmers considering such a decision would take into account a number of other changes as well. Although the improved cotton technology slightly increased the returns to labor relative to cotton grown with indigenous technology, farming families would also find that:

1. They would sacrifice some returns to labor, particularly during the June-July period, when food crops would yield a better return per man-hour (Table 7.4).
2. There would be increased inter-crop competition for weeding time if both food crops and cotton had to be weeded during the June-July labor bottleneck because some hand weeding, even with oxen cultivation, would still be required (Figure 7.1).

Because of those factors, recommendations were developed for July-planted rather than June-planted cotton. Although the potential yields of late-planted cotton would be lower than the early planted, they still would be potentially higher than yields using indigenous practices and such an adjustment would fit in better with current farming systems and the household profit expectations.

For farmers facing land limitations, the results of the testing indicated that, on the average, the yields and profitability per hectare of the improved technologies for all crops were substantially higher than for the same crops grown under indigenous conditions.[14] They also indicated that, on average, the improved sorghum technology was considerably more profitable than that for

TABLE 7.4
Average Inputs and Returns from Improved Technological Packages, Daudawa and Hanwa[a]

| Variable Specification | Oxen Power | | | | | Hand Power |
| | Cotton | | Sorghum | | Maize | Cowpeas |
	Indigenous	Improved	Indigenous	Improved	Improved	Improved
Inputs (per ha.):						
Fertilizer (N:P:K)[b]:	1:0:0	27:22:0	0:0:0	95:46:0	189:49:49	0:0:0
Labor (man-hours)[b]:						
Total	276	430	199	337	354	718
June-July	55	110	46	100	107	200
Harvesting	124	221	102	196	214	328
Costs (N/ha):						
Non-labor costs	9.22	31.00	11.71	40.92	65.90	56.01
Labor[c]	23.41 (60)	36.07 (76)	20.53 (56)	34.72 (62)	36.32 (55)	69.53 (56)
Net return:						
N per ha	19.63	40.73	45.01	81.62	190.36	60.73
N per man-hour[d]:						
Total	0.15	0.17	0.32	0.33	0.62	0.25
June-July[e]	0.33 (1.67)	0.38 (1.41)	1.02	0.84	1.78	0.38
Excluding harvesting	0.20	0.25	0.54	0.63	1.39	0.29

[a]All figures in the table represent averages for the years in which the technological packages were tested (see Table 7.2).

[b]Excludes threshing (except for cowpeas) and time spent travelling to and from the field.

[c]Includes imputed value for labor. Each figure in parentheses represents the percentage of labor hired.

[d]The figure is calculated by subtracting from value of production (N/ha) the sum of nonlabor costs (N/ha) and total labor costs excluding labor in denominator (N/ha) times the opportunity cost of capital (assumed to be 12 percent) all divided by the man-hours in the denominator (i.e., total man-hours, June-July man-hours or man-hours excluding harvest).

[e]Because under indigenous conditions, planting of cotton was done in July, the figures in parentheses express the return per man-hour of labor put in during June.

FIGURE 7.1
Monthly Composition of Work on Cotton, Daudawa, 1972

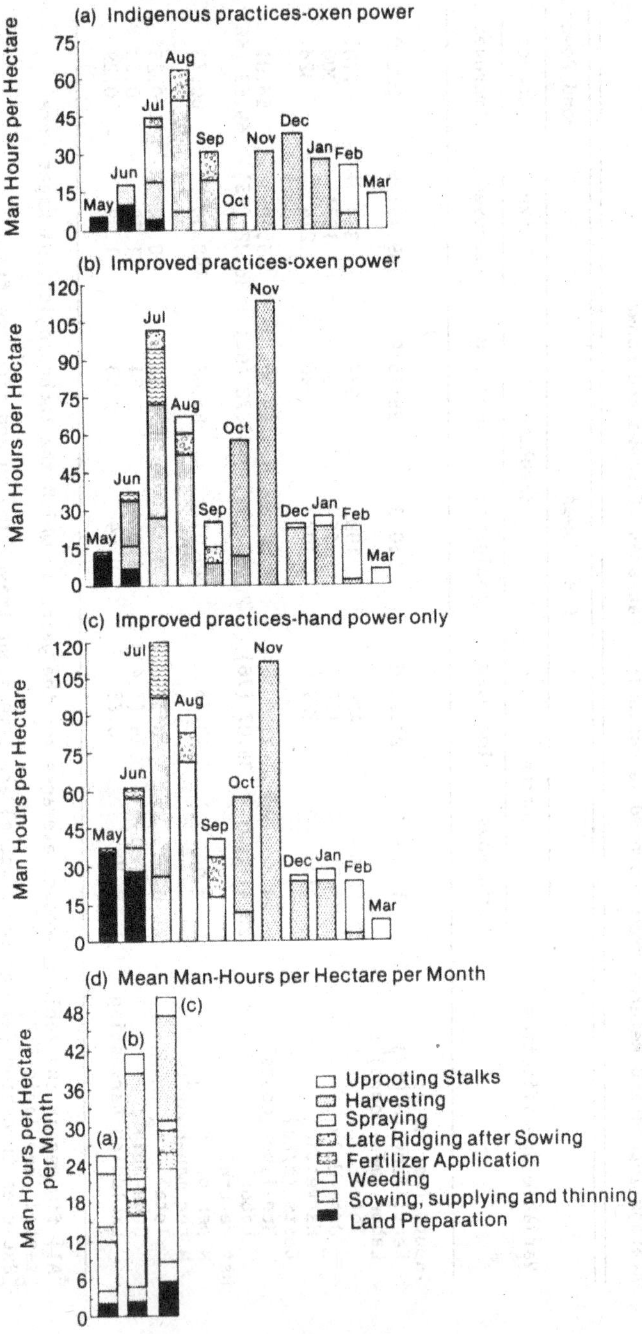
(a) Indigenous practices-oxen power
Man Hours per Hectare
75
60
45
30
15
0
May
Jun
Jul
Aug
Sep
Oct
Nov
Dec
Jan
Feb
Mar
(b) Improved practices-oxen power
Man Hours per Hectare
120
105
90
75
60
45
30
15
0
May
Jun
Jul
Aug
Sep
Oct
Nov
Dec
Jan
Feb
Mar
(c) Improved practices-hand power only
Man Hours per Hectare
120
105
90
75
60
45
30
15
0
May
Jun
Jul
Aug
Sep
Oct
Nov
Dec
Jan
Feb
Mar
(d) Mean Man-Hours per Hectare per Month
Man Hours per Hectare per Month
48
42
36
30
24
18
12
6
0
(a)
(b)
(c)
Uprooting Stalks
Harvesting
Spraying
Late Ridging after Sowing
Fertilizer Application
Weeding
Sowing, supplying and thinning
Land Preparation

improved cotton, even though neither compared favorably with improved cowpeas in 1974 or with improved maize in 1973 or 1974.

The impact of improved technology on the returns per unit of labor was mixed. In all cases, the improved technology packages required substantial increases in labor inputs, with a marked shift in the distribution of labor. An average of 65 percent of the extra labor was devoted to harvesting the additional yield. The general implication of this increased labor demand for women's seclusion has already been noted, but it also implied that some household adjustments might have been required. When the absolute increase in labor requirements and the changed distribution were considered, the improved technology for sorghum was less promising than the indigenous technology, in terms of returns per June-July man-hour, but the cotton package was even less so (Table 7.4). Both resulted in insignificant increases in return per man-hour overall.

Although oxen could substitute for only some of the hand labor, use of oxen did increase the average returns to labor. That differential was augmented further when oxen power was combined with the improved-technology packages for sorghum, maize, and cotton (Norman, Pryor, and Gibbs 1979).[15] Because most farmers did not own carts, however, the introduction of oxen tended to accentuate the harvesting bottleneck relative to the planting and weeding bottlenecks. Figure 7.1 illustrates that for cotton.

Coverage of cash costs turned out not to be a problem. The coefficients of variation indicated that there was little difference in relative risk for indigenous and improved technological packages, and yields were (except for cotton in the drought year of 1973 and for cowpeas attacked by insects in 1975) in excess of those needed to cover all costs of production (Table 7.3). Such dependability of return is very important in ensuring that cash risks assumed for adopting of improved technologies are within the farm families' capacity. It should be emphasized, however, that credit and market factors were not a problem in these tests. Wider production of maize, for example, could be severely undermined if some attention is not directed toward market development or support pricing in the initial stages of establishing the crop in the area.

Lessons from Experience

From our own experiences in working with farmers, we learned several valuable lessons that helped us formulate our convictions on the potential significance of a farming systems approach to research. We here discuss five of the major ones.

Interdisciplinary cooperation at the farm level. By working at the farm level we obtained a much better understanding of the interaction between the technical and human elements of the environment. As a result we were sensitized to the fact that there could be a degree of location and farmer specificity in determining relevant, improved technologies and support systems. That in turn convinced us of the validity of replacing the common top-down approach with the bottom-up approach characteristic of FSR and of the necessity of a multidisciplinary team working in an interdisciplinary manner, with the social scientist playing an ex ante constructive role rather than the more traditional ex post role. For example, in the case of the cotton and the cowpea packages, it was not necessary for us to confront our co-workers from the technical sciences with the problems of the packages. By working together with us they were immediately able to perceive the problems themselves. Therefore, our role was never perceived as being destructive; instead, we were able to work together to devise possible solutions to identified problems. At the same time, however, we found that working in an interdisciplinary team was not always easy. To be able to work together effectively, team members had to have other characteristics besides compatible personalities. Each team member had to be convinced that he/she could contribute constructively to the work and had to be confident about the role of his/her own discipline. At the same time, however, he/she needed to appreciate the limitations of his/her own discipline in solving the problems of farming families without complementary inputs from other disciplines. Unfortunately, it appears that this disciplinary maturity currently comes mainly from longevity in the field rather than through formal training programs. Finally, each team member must, when necessary, be prepared to undertake tasks that are in the team's interest but that fall outside the mandate of his/her own discipline.

Central role of the farming family. We became convinced that in the research process the farmer must be the central figure and that including farming households increases the probability of developing improved systems that will address the constraints the households face, will recognize the multiple uses for their productive resources, and will be evaluated in terms relevant to them. In addition, including them makes it possible to use their intimate local knowledge. For example, the good elements of the systems they currently practice (such as using certain crop mixtures) can be incorporated into the proposed improved system. Critically important in determining the quality of the contribution that farmers make is the strength of the interactive process between them and researchers. The relationship has to be based on mutual respect. Too often research workers tend to be paternalistic with farmers who have little or no formal education, giving demonstrations and telling farmers what to do. With that approach, two critical components of FSAR are lost: two-way dialogue involving both talking and listening on the part of research workers, and the concept of testing under practical farming conditions. We of course recognize that not all farmers, like not all research workers, can articulate their thoughts well, but we have had enough useful learning experiences from farmers to convince us that valuable insights often can be obtained from listening. Although many farmers have responded to queries as to the "why" of particular agricultural practices by citing "tradition" or "my father showed me how to do it," increasingly we have come to the conclusion that certain traditional practices have continued to the present because they have enabled successive generations to survive, and many are based on scientific principles. On the other hand, some farmers have surprised us with their well-articulated statements. One farmer, for example, essentially told us that he had two important guidelines for growing crops in mixtures: first, that the different crops should have complementary growth cycles; and, second, that certain crops, such as cowpeas, have beneficial effects on other crops! That does not mean that we advise researchers to believe and accept everything farmers say. The validity of their comments must be evaluated, as must those of other researchers.

In our testing, we found a technical package that did not work well to be no problem, provided the farmer was well aware at the outset that it was a test and not a demonstration. If the return from an improved-technology package is in some doubt, then farmer-managed tests should be preceded by researcher-managed tests in which some guaranteed level of return can be promised to participating farmers if the tests fail.

Emphasis on the testing stage. We became fully convinced that work at the farm level should receive greater emphasis than it is now receiving and that such testing should be conducted for more than one year before the technology package is released to the extension service for dissemination. The cowpea package illustrated the desirability of both farm-level and multi-year testing, during which both researcher-managed trials and farmer-managed tests have important roles. In the farmer-managed tests, we found that it is essential to establish plots large enough to collect valid labor-flow and yield data. At that level of testing, because of the nature and size of such plots, only one suggested package of practices per farmer is usually possible, and replications have to occur between fields--usually farmed by different farming families--rather than within fields.

Dynamic and iterative nature of the research process. We came to appreciate how important it is for research procedure to be a dynamic, iterative process as well as to appreciate the complementary nature of the relationship between farm-level research and experiment station-based research from which the body of knowledge is created (Figure 2.2). The problems of water for spraying and early planting of cotton illustrate the critical value of such a linkage and relationship.

Fallback strategies and the research-extension interface. Through our relationship with the extension staff, we came to two conclusions: first, that both farmers and extension staff face problems with the package approach; and second that, under the umbrella of FSAR, it is essential that researchers and extension personnel work together at the farm level, probably necessitating some inter-institutional agreement. With regard to the first conclusion, if farmers failed for one reason or another to apply the specified quantity of fertilizer or did not apply it at the right time, extension agents were not in a position to suggest fallback strategies for the other inputs and operations. We became convinced that researchers need to be able to suggest possible fallback strategies to improve the effectiveness of extension efforts.

TESTING DIFFERENT SUPPORT-SYSTEM DESIGNS

The guided change project (GCP) represented an attempt to assess the effects of alternative means of introducing change into villages. The GCP provided support systems on a pilot-project basis in a traditional agricultural setting in Giwa District near Zaria. The project had as its basic purposes (Huizinga n.d.):

1. To develop a number of alternative prototypes for action at the village level, aimed at bringing about changes in the status of agricultural development.
2. To assist government in implementing those prototypes by helping to remove constraints in their implementation.
3. To evaluate the prototypes in terms of their effectiveness in bringing about agricultural development.

The project involved testing three approaches to facilitating farmer use of well-tested packages of agricultural technology developed at IAR. Three groups of four villages each were selected for trials of the three treatments:

1. Cash villages, where an input-distribution program was introduced and farmers were given the opportunity to buy the inputs for cash.
2. Credit villages, where the distribution program for inputs gave farmers the opportunity to buy the technology packages on credit.
3. Extension villages, where the input packages, credit, and an extension service were introduced simultaneously.

The GCP in its 4 year life reached some 4,000 farming families each year and involved costs of approximately ₦575 per village per year. In addition, during one year all villages were reached by a weekly five-minute radio program providing information on the operation of the project and on times for collecting inputs, making credit payment, and doing other tasks.

The GCP was largely successful in attaining its first and third research objectives: giving credit and expanding the extension input encouraged the greatest use of inputs; instituting a creative, villager-managed savings program with the credit program enabled farmers to pay back loans on time and thus reduced defaults and administrative costs involved in multiple-collection trips; and providing for greater use of the inputs was associated with higher yields. The GCP team also found that ensuring all farming families equal access to project resources was important in generating increased overall as well as individual production benefits and that individual rather than group-credit programs were the only way to achieve such access and, incidentally, to maximize repayment.[16] On the other hand, the GCP was not successful in building government capacity to replicate the project's findings; indeed, in the government-sponsored Agricultural Development Project in adjacent Funtua District, input distribution was implemented by using approaches the GCP had just begun to demonstrate were likely to be less effective.

Nevertheless, the GCP illustrated that, with an appropriate mandate, an FSAR program could contribute to technology dissemination as well as to technology development.

INCORPORATING ON-FARM TESTING IN THE FARMING SYSTEMS RESEARCH APPROACH

As we have just demonstrated, testing at the farm and village level provided a realistic environment for evaluating the potential suitability of proposed improved technologies, techniques, and support systems. Usually, for example, the performance of improved technology drops when it moves from the somewhat artificial conditions of the experiment station to trials managed by researchers at the farm level and drops again at the farmer's testing level when the improved technology is, in effect, being tested for compatibility with the current farming system, the managerial know-how of the farmer, and the adequacy of the support systems needed to facilitate farmer adoption. It is important to note several distinctions between the two types of on-farm trials.

Researcher managed tests (RMTs) involve heavier inputs of management from the researchers than do farmer-managed tests (FMTs) so less independent participation is expected from the farmer. Experimental designs, similar to those for trials on experiment stations, can be used for RMTs, which can include more treatments and replications than those managed by farmers themselves. In using RMTs, the aim is to screen the improved technologies arising from the design stage, to fine-tune them to the local situation, and to evaluate their potential both locally and for broader regional coverage. RMTs can consist of replications either within fields or between fields, to check on-site variability.

But farmer-managed tests or trials (FMTs) provide the more rigorous tests of the proposed improved technologies. FMTs generally involve treatments that are less complex, but their performance criteria are generally more complex. Whereas RMTs tend to focus relatively more on issues of technical feasibility, FMTs take into account the full range of suitability criteria that we have just discussed. The field treatments are more simplified not only because of cost, but also in order to facilitate meaningful interaction between farmers and research workers. Such interaction is more difficult to achieve when the experiments are very complex and involve many treatments.

A critical issue that often arises at the testing stage is whether to maximize the chance for a good interactive process between farmers and research workers or to insist that a broad range of farmers and farming families be represented. Some researchers prefer to select the better, more responsive, or more cooperative farmers to participate in the testing stage. Using cooperativeness as a criterion has the advantage of improving communications between researchers and farmers. But there is the potential problem that even when improved practices receive a positive evaluation, they may not be truly relevant for other groups of farmers and farming households. The adoption process might thus be biased toward farmers with particular characteristics and could cause inequalities in benefits of distribution in the long run. Other research workers in FSR, therefore, advocate selecting a cross-section of farmers representative of the subgroup or subgroups for whom the technologies are thought to be appropriate. The possible disadvantage, that selecting representative farmers might not maximize interactions between farmers and research workers, would be offset by the big advantage of getting a more satisfactory idea of whether the improved practices would be suitable for the average farmer.

Three other points, already alluded to, need to be considered in designing FMTs to provide valid, useful data for evaluating improved practices:

1. As we indicated earlier, it is important that the plots be large enough to accommodate the improved technologies being tested. Labor is an important input, and to farmers the returns per labor-hour are an important criterion of performance. Plots need to be large enough for labor inputs to be accurately measured. Consequently, replications within the field are seldom possible. The improved technology, however, could be replicated on fields of other farmers.

2. Both technical and human environments vary widely over time. Testing for more than one year gives a better idea of the level and stability of the improved practices, particularly if inter-annual variations in the "total" environment are substantial. In effect, replications can be increased by incorporating the time dimension, using the same improved practices in different years. But such a replication objective should not preclude modifying of the tests if results were unsatisfactory in the earlier years.

3. To provide valid evaluation of improved practices, it is important to obtain data that can be used to assess compatibility of the practices with other parts of the farming systems. Two alternative approaches might be used: collecting data on all parts of the farming system to assess potential conflicts and compatibility; or (the one more often adopted) minimizing costs of data collection by focussing only on the parts of the farming system that the improved practices are likely to affect directly or to replace. But that should be done cautiously if adopting the improved practices requires a significant change in the flow and level of resources. An example of such a change in our test of technology packages was the planting date of cotton.

Successful farmer-managed trials obviously would indicate the potential for wider replicability of the improved technology packages and would be followed by extension service training and extension campaigns to promote the technologies. That might seem to signal the end of the researcher involvement, but continued assessment at this stage would provide information on changes taking place in the agricultural sector and on farms and would help to identify new problems--second generation problems--of logical concern for a new iteration of the farming systems research approach. Monitoring and evaluating extension efforts would check the validity of the descriptive/diagnostic, design, and testing activities of the FSR, so lessons could be incorporated systematically into future FSR programs. Such monitoring and evaluation also could indicate to the researchers the need to rally the pressure for changes in some of the exogenous variables--such as policies, prices, and infrastructure. Using the more direct approach (such as that used in the GCP) to test those variables might be one way of considering these questions. Although the results of the approach were potentially useful, the project also illustrated some of the organizational difficulties that can emerge when the "sponsoring agency" has been so substantially extended.[17]

Monitoring and evaluating the introduction of improved strategies need to be examined from the perspectives of researchers, farming families, and society as a whole. The research perspective is reflected in the degree to which the needs of the individual farming family and society are met. In monitoring, it is important to determine how many individual farming families have adopted the improved technology, the degree to which they have adopted it (including the different components of a package), and the reasons for divergence from what was recommended. Some types of information necessitate acceptability-testing procedures. Acceptability or adoption indices like those suggested by Hildebrand (1979a) can be a valuable aid. Evaluating the impact of improved technology from the viewpoint of society involves answering such questions as the distribution of benefits from its adoption, stability of the ecological base, and the general nutritional level.

LOCATING A "FARMING SYSTEMS APPROACH TO RESEARCH" PROGRAM

Prospects for successfully introducing FSAR programs at the national level are influenced by a complex of intra- and inter-institutional relationships involving national agricultural institutions and universities; implementing agencies, including Ministries of Agriculture and Natural Resources, and of Rural Development; planning departments; and funding agencies. Most FSAR programs involving diagnosis, design, and on-farm testing such as that described earlier in this chapter are commonly and logically associated with agricultural research institutions. FSAR activities may not, however, be readily accepted by such institutions for several reasons:

1. Resource limitations. National agricultural research organizations in developing countries are generally thinly staffed, sometimes include a high percentage of expatriates, are poorly supported, and depend heavily on external donor agencies for assistance--often even for some recurrent expenses. Such organizations often hesitate to initiate FSAR programs on their own account because doing so diverts resources from resource-starved, on-going component research.

2. Reluctance to change. Most scientists at agricultural research institutions have been trained in and have experience in disciplinary and commodity-research programs, so many have limited understanding of and mixed feelings

about FSAR. Research institutions also are normally set up along disciplinary or commodity lines, so incorporating FSAR can create jurisdictional problems and present formidable obstacles to redefining responsibilities. In addition, social scientists commonly are neither found at nor particularly welcomed in agricultural research institutions.

3. Self-sufficiency and professional image. People in many developing countries resist looking to outside regional or international institutions for research results that can be adapted to local situations. Many think that borrowing technology will relegate the in-country research establishments to permanent secondary or even tertiary status in the hierarchy of agricultural research.

4. Time required to establish an efficient and credible FSAR program. Even where existing agricultural research institutions agree to initiate FSAR-type activities, they may not have the patience to allow the activities to become effective. Researchers charged with implementing FSAR programs characteristically have little or no experience in interdisciplinary team efforts. An FSAR team gains experience and credibility over time and through the continuity of staff. Further, linkages with planning, funding, and implementing institutions also take time to develop.

We view FSAR as a process that generally can be either incorporated into existing research programs as a philosophy of research or established as a separate administrative and substantive unit within an agricultural research institute. It is not necessary, nor perhaps even desirable in many instances, to have an administratively independent farming systems research unit. Several agricultural research institutes in developing countries already have quasi-FSAR activities that simply evolved from collaborative projects or from a tradition of on-farm trials--generally of a researcher-managed type. Such an evolutionary process might be the most effective way of promoting the farming systems approach to research, even for an activity not labelled FSAR. Of course in some situations, such an evolution might never emerge. When agricultural research and development policies are not focussed on the needs of small farmers but are, for example, geared strictly to increased production of an exportable crop, FSAR might take root only as part of a general orientation and reorganization of the total research system. That usually would presuppose a national government decision to rethink the objectives and focus of its agricultural research program.[18]

Lack of apparent productive impact might be one reason for such a policy, though FSAR programs could come about for other reasons as well. National agricultural development banks and donor agencies could be potential allies of FSAR-type activities at the national level. Those agencies have procedures for identifying, designing, appraising, monitoring, and evaluating rural/agricultural development projects. In addition, they often have policies that explicitly direct them to devote a substantial, if not a major, share of their resources to assisting rural areas and the rural poor. In many cases they also actively seek ways to improve upon the somewhat mediocre performance of their efforts, particularly where improved technologies that small farmers can use have proved elusive. Some of those agencies, however, are staffed with veterans of agricultural development who contend that implementing the farming systems approach to research will be too complicated, costly, and time-consuming to be useful. Preparing a project is a lengthy, involved process; therefore, they perceive developing a new farming systems approach to research as another bottleneck to project implementation. If the potential advantages of the FSAR are readily apparent, however, this issue becomes somewhat of a red herring.

FSAR programs may have an ally in national and regional planning agencies. Many planning agencies are poorly staffed and not effectively integrated into governmental decision-making processes. Yet they often are given responsibility for approving development projects and generally assessing the merits of annual budgets. That makes them receptive to mechanisms that can improve project designs and assist them in monitoring/evaluating on-going projects. To require all implementing agencies to use FSAR in the first instance, however, might only create serious bottlenecks, because the capacity to provide such services is not likely to exist in most countries. A particular problem is a lack of individuals trained and experienced in FSAR. So a gradual, selective introduction of the farming systems perspective is probably preferable.

In summary, a range of inter- and intra-institutional issues at the national level bear directly on the feasibility of the farming systems approach to research as a means of developing relevant, improved agricultural techniques and facilitating their adoption. Resolving the institutional issues is one of the keys to FSAR's future utility. Ironically, the conditions that have made increasing numbers of institutions look to FSAR as a way to improve agricultural development in specific locations mitigate against achieving a spectacular Green Revolution-type of breakthrough for large areas that would give great impetus to the development and acceptance of FSAR. The spectacular breakthroughs that took place in the relatively few well-endowed areas of the developing world--such as the Punjab--are not likely to be repeated in less favored areas where smaller incremental changes are more likely. In addition, FSAR is by nature conservative because it is linked to helping farmers in the context of existing farming systems.

IMPROVED TECHNOLOGIES AND SUPPORT SYSTEMS

In earlier chapters we showed that there is considerable heterogeneity in the farming systems practiced by farming families in the West African savanna. We also presented some evidence to indicate that this heterogeneity might be growing. In this chapter we have argued that FSAR can help in introducing a degree of specificity into the design of relevant improved technologies and support systems. The specificity required will depend on the location and particular group of farming families whose needs are being addressed.

In general, recognizing differences and trends in population density helps bring into focus three problems that exist in different parts of the savanna:

1. In areas of low population density, the peak demand period for labor is likely to be a major constraining factor on expanded output.
2. In areas of transition to high population densities, it is possible that both a labor and a land constraint will emerge. The peak demand period for labor will be a constraining influence and land will emerge as a problem because soil fertility will decline under population pressure. The possible dual nature of these constraints will be exacerbated by the increasing necessity for farm families to spend more time in activities that require year-round commitment, including off-farm income-earning activities, as well as caring for cattle owned by the family. As land becomes more of a constraint, the value of cattle in contributing to maintaining soil fertility will become greater. However, the problem of feeding the livestock also will become greater and quite likely will involve a change to more labor-intensive methods.
3. In areas of very high population density, where labor becomes surplus, land is likely to be the most constraining factor.

With the inability of the nonagricultural sector to absorb the substantial increases in population, it is likely that scenarios two and three will become of increasing significance. That trend could be exacerbated if inequalities in land distribution increases which we earlier suggested might in fact now be occurring in conjunction with the move toward increasing interaction with the economy and society outside the villages.

To date the constraints articulated here have been largely overcome within the traditional farming system framework. Crop diversification--involving use of a crop-mixture strategy--and the adoption of various ways of increasing the labor input on the family farm are being used to overcome the problems of the labor-bottleneck period. Raising cattle and seeking off-farm occupations are being used to combat the problem of decreasing soil fertility. We suggest, however, there is limited potential for continuing to overcome these problems by using indigenous solutions. Consequently, if nothing is done to lessen these constraints, involution likely will occur.[19] Also, even if the requisite incentives were present, the low productivity of both land and labor under such systems likely will not permit the generation of sufficient surplus food production to feed the rapidly increasing urban population.

The future, therefore, has to lie with developing and introducing relevant, improved technology. The scenario of problems can be reduced to two basic constraints, with relative significance depending to a large extent on the seasonality of agriculture and population pressure: first, improving the productivity of labor, particularly at bottleneck periods; and second, improving the productivity of land on a sustainable basis. Improved technology development needs to address these issues in order to increase the productivity of the existing farming systems. Concluding that mechanization can be used to solve the problems of seasonal labor bottlenecks, and bio-chemical technology to increase land productivity, however, is too simplistic. As well as these direct effects such technology changes would have important indirect effects. Deep plowing with mechanical equipment, for example, was earlier mentioned as a possible way to sustain land productivity in the long-run.

From an economic viewpoint Hayami and Ruttan (1971)[20] emphasized in their induced-innovation hypothesis that, although to increase productivity it is necessary to increase the return to the most limiting factor, that action alone will indirectly affect the use and productivity of other inputs (Table 7.5).

TABLE 7.5
Relationship Between Types of Required Technology and Land/Labor Ratios

Land/Labor Ratio	Technology Required	Productivity of[a]	
		Land	Labor
High	Labor saving	I + or -	D +
Low	Yield increasing	D +	I + or -

[a]D = direct impact
I = indirect impact
+ = positive impact
- = negative impact

In summary, both developing improved technologies and evaluating their relevancy are complex matters. Figure 7.2 demonstrates graphically both the dimensions of the problem and the difficulty of the tasks in many parts of the West African savanna. The schematic diagram arrays along different axes five interwoven variables: household goals, market and support system development, population density, market opportunities, and primary technology development requirements. Several conclusions are implict in the schema:

1. Population density is important; it effects the technology emphasis. In areas of low population density (areas 1 and 4) labor-saving strategies are more significant whereas in areas of high population density (areas 3 and 6) yield-increasing strategies are required. At intermediate levels of population density (areas 2 and 5) both technological options must be taken into account.
2. Market system development permits household goals to be redefined. With the development of market systems, it is potentially possible for the traditionally important goal of

FIGURE 7.2
Schematic Breakdown of Relationship Between
Population Density and Market-System Development

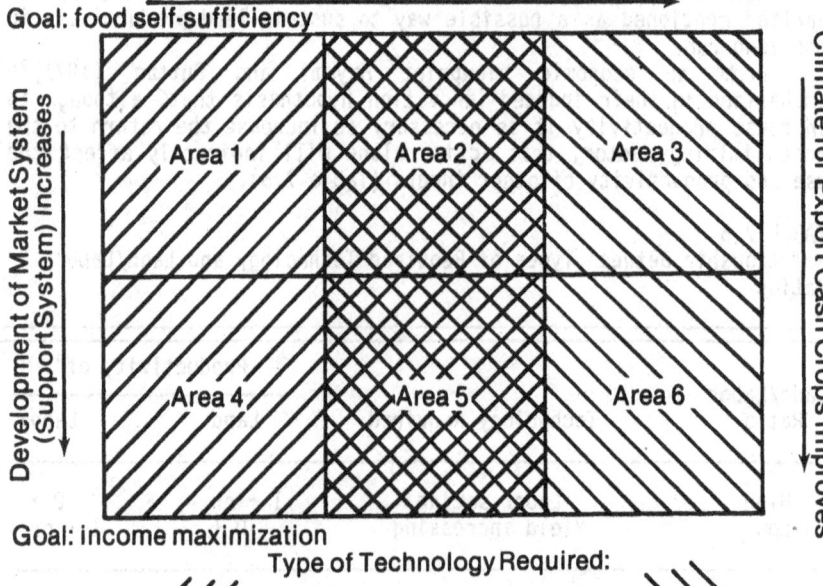

Land/Labor Ratio
Decreases (Population Density Increases)

Goal: food self-sufficiency

Development of Market System
(Support System) Increases

Area 1   Area 2   Area 3

Area 4   Area 5   Area 6

Climate for Export Cash Crops Improves

Goal: income maximization

Type of Technology Required:

/// Labor Saving   Yield Increasing \\\

food self-sufficiency to become at least partially diluted
in favor of a more commercialized agriculture that involves
entering the market place. In general, however, the history
of market-system development in the West African savanna
shows that developing markets for improved inputs and
input-related services has lagged behind those relating to
the product-marketing side. Therefore, the introduction of
improved technologies for crops may be slowed, particularly
in areas where market systems for improved inputs and input
related services are still relatively poorly developed
(areas 1 to 3). As we mentioned earlier, historically there
has been a bias in market-system development particularly on
the input side toward those areas where rainfall is high
enough for export cash crops to be grown (areas 4 to 6).

3. While the ability of oxen to substitute for labor is an
obvious attraction, cost factors affect their potential in
the West African savanna. Where new inputs are part of a
technology package, their ability to substitute for or to
complement other inputs must be considered. Because oxen
are not part of the current food production sector,
labor-saving technologies involving animal traction have

worked better where rainfall is adequate to allow export cash crops (areas 4 and 5) rather than where market system development is poor (areas 1 and 2). Also, they have worked better when combined with yield-increasing technologies intrinsically more relevant to an area like 6.

4. Input delivery systems and input-related services are likely to be more relevant in areas where land is a constraint. Yield-increasing technologies including use of improved seed, fertilizer, and pesticides, which are the primary focus of most technical scientists, make it easier for scientists to develop relevant improved technologies suitable for an area like 6 than for one like 4. Also with scientists' current orientation, prospects are not good for developing technologies that will benefit farming families in areas such as 1 to 3, where market systems are generally poorly developed. Greater relevancy, although perhaps not with spectacular increments in productivity, could be achieved in such areas through scientists changing their orientation from modifying the environment to fit the plant, to modifying the plant to fit the environment.

5. The interventionist approach to market system development is critical in ecologically fragile areas. It is impossible to expect major technological breakthroughs in ecologically fragile areas such as 3 without substantial inputs from outside agriculture. Thus, in such an area some emphasis will have to be placed on developing improved technologies that require market-structure development and, through their potential, on providing the pressure for that development. It is in areas 2 and 3 that the greatest challenges lie; not only are marketing systems poorly developed there but also the unexploited carrying capacity of the land is low compared with that in areas 5 and 6.

Although we have here discussed the interdependency between support systems (market-structure development) and improved technologies from an inter-areal perspective, earlier we discussed interdependencies that exist within villages between the two types of factors. Within communities it is obviously desirable to design and implement strategies that will help all farming families. Such strategies involve designing relevant improved technologies and support systems. Heterogeneity within the villages must be recognized in designing such strategies. The challenge is to find ways to help the disadvantaged farming families. It is easy, for example, to design improved technologies suitable just for large-scale farmers but it is almost impossible to design improved technologies that are suitable only for small-scale farmers. Another problem results from accessibility to support systems. Where they are limited or there is a hierarchical village-authority structure, the probability of differential access is greater than elsewhere. The problem in such situations is to design a cost-efficient support system that will ensure equitability of access and at the same time will not alienate the village leadership.

NOTES

1. Abalu and Harkness (1976) were involved in testing an improved technology package for groundnuts, not analyzed here. Additional details on the various packages examined in this chapter are given in Beeden et al. (1976), Norman et al. (1976a and 1976b), and Hays and Raheja (1977).

2. The general practice in fact was to release the recommendations to the extension service, which disseminated them to farmers in part by using them on demonstration plots. That provided essentially the only formalized data resulting from the packages on farmers' fields, and in any case these fields were largely extension managed.

3. In the 1966-67 study in the Zaria villages, where only hand labor was used, the percentages of total adjusted hectares of sorghum, maize, cowpeas, and cotton grown as sole stands were 27, 27, 2, and 31, respectively (Norman 1972).

4. The higher densities were achieved by decreasing distances between plant stands within rows rather than between rows.

5. Seed and seed dressing are provided free. Reissue of the seed each year is necessary because the lint has to be removed from the seed at a ginnery before it can be used.

6. These farming families were on a settlement scheme: the authorities had originally aided them in obtaining oxen.

7. The maize seed purchased by IAR made it possible to have more seed to distribute to more farmers the following year.

8. For example, seven of the ten test farmers were in the 1966-67 farm-management study. In 1966-67 these were the respective averages for the test farmers and the overall Hanwa average: 10.4 and 10.9 for size of family; 3.6 and 2.9 hectares for size of farm, ₦233.69 and ₦215.43 for family disposable income; and ₦21.74 and ₦22.08 for disposable income per resident. The disposable income figures excluded that derived from cattle. Four of the seven farming families owned cattle.

9. In fairness to IAR, the recommending agency, it should be noted that the area in which this variety was tested was slightly north of the recommended zone.

10. This was consistent with the lexicographic behavior pattern suggested by Balcet and Candler (1981), which we mentioned in Chapter 6.

11. In fact, because of these problems, the spraying operation was done under contract, the farmers paying for the service at harvest time. A contract operation enabled a motorized knapsack mistblower to be used; it required only 135 liters of water per hectare per spray. When in later years the switch was made to ultra-low volume sprayers farming families started doing the spraying operation themselves.

12. There were, however, promising indications that such a market could develop. The test farmers in both years of the maize testing kept approximately 40 percent of their production for consumption and gifts. That amounted to withholding from sales 501 kg per family in 1973 and 667 kg per family in 1974. These figures imply that maize had become a large consumption item in those households; they compared with 42 kg per family consumed per year in

the consumption study in the Zaria area (Simmons 1976a). In another survey, in Daudawa, about half of the farming families indicated they would be willing to consume more maize (Awolola and Buntjer 1976).

13. Despite some pressure from farming families for cash credit to pay for nonfamily labor, that was not considered feasible.

14. This did not apply for cotton in 1973. The big increase in profit for cotton in 1974 was due not only to the improved yield but also to the substantially higher price set by the marketing board for seed cotton.

15. These results provide support for the observation (see Chapter 6) that oxen have been most successfully introduced where land-intensive technologies--particularly for export cash crops--have been widely adopted (for example, Mali Sud in Mali and Sine Saloum in Senegal).

16. Unlike the type of group credit program used to facilitate the testing of the technology packages in Daudawa and Hanwa, the type of group credit program referred to here involved both extending and repaying of credit through groups.

17. Although substantial help was received from the Kaduna State Ministry of Agriculture and Natural Resources, much time was spent by IAR researchers in implementing the project. Some perceived the high proportion spent on such activities as being an incorrect use of time for staff of a research institute. Unfortunately, of course, the test or research component could not be undertaken until the project had been implemented!

18. FSAR-type activities were in fact initiated in Guatemala after a major reorganization of the national agricultural research system (Fumagalli and Waugh 1977).

19. Involution means a higher total income in an area but because of population increases, a lower income per capita.

20. More recent definitive work in this area has been done by Binswanger and Ruttan (1978).

8
Promoting
Agricultural Development

"The five national objectives of Nigeria . . . are to
establish Nigeria firmly as a united and self-reliant
nation, a great and dynamic economy, a just and
egalitarian society, a land of bright and full
opportunities for all citizens, and a free and democratic
society . . . An important objective of the Plan,
therefore, is to spread the benefits of economic
development so that the average Nigerian will experience a
marked improvement in his standard of living."
                    Federal Republic of Nigeria (1975)

Given the complexities and variability in savanna farming
systems just described, the task of promoting agricultural change
and improving welfare appears to be difficult indeed. Adopting
simplified assumptions about farm-level constraints could, as we
have shown, lead to the development of agricultural technology that
is not relevant to the majority of farming systems. Or it could
lead to expectations of response to policy changes that are not
warranted. On the other hand, excessive caution about farmers'
capacity to take on risks and to experiment with new technologies
may be equally misplaced. Population totals and the demand for food
continue to grow, and productivity of the agricultural resources in
the savanna nations of West Africa must be increased.

Nigeria is not alone among the savanna states in its efforts to
implement programs that will promote the needed agricultural
development. Nigeria possesses, however, a more substantial volume
of resources--both financial and human--to put behind its
intentions. As we stated at the outset, starting with a "micro"
perspective such as that embodied in the farming systems approach to
research permits the formation of agricultural policies and
strategies tuned to the incentive structures and resources of the
producers themselves--making it possible to anticipate and
ameliorate conflicts between national goals and farmers' goals
before problems become apparent and tensions arise. With
appropriate attention to institutional and methodological issues, a
farming systems research program can, therefore, be useful in
designing agricultural strategies that address farm-level
constraints directly and cost-effectively. Where broad, indirect
strategies are relied upon to promote agricultural growth, farming

systems approaches to research are likely to be most appropriately applied to the level of policy analysis. A farming systems perspective may improve the chances for policies that only indirectly affect agriculture to have the desired impacts on the sector. For nations that have adopted, because of the urgency of food-production needs, policies advocating production approaches that bypass, rather than include, the small farm sector, the farming systems approach to research can offer little assistance.

INDIRECT STRATEGIES: EXPLOITING UNUSED RESOURCES

Before Nigeria became independent, the success of strategies for development lay in the ability of the country's farming households to put new lands under cultivation and to mobilize more household labor in response to the general efforts of the colonial governments to provide better access to international markets. As Helleiner (1966) pointed out, normal market incentives combined with a substantial amount of transportation infrastructure development were sufficient to induce a growth in agricultural output without encouraging the broader adoption of improved technologies. Farmers in the north of Nigeria produced cotton and groundnuts in much the same way as they produced traditional food crops: using no pesticides, no fertilizers, and no machinery.

Expansion of transportation networks and modifications of market prices are still viewed by post-colonial governments of West Africa as powerful, indirect approaches to agricultural growth. The rationale behind those strategies lies not only in the lessons of history, but also in the perception that there is currently considerable underuse of production resources, primarily land and labor.

In many areas, as in our isolated village with substantial areas of under-exploited fadama, an assumption that improved transportation alone would be sufficient to induce additional agricultural growth may be correct. In other areas, however, there is considerable evidence that seasonal labor bottlenecks already have limited the amount of land expansion that could be undertaken by farming families, that scarcity and variability of rainfall now limit the additional risks that farmers would be willing to undertake, and that soil fertility already may have declined under intensified population pressure on land to the extent that additional inputs of labor would yield low returns. Thus, a farming systems perspective on the impact of policies designed primarily to permit further exploitation of unused or under-utilized resources, without additional inputs or improved agricultural technologies, indicates that strategies that rely primarily on this mechanism to spur agricultural development may be more limited and more location-specific than anticipated.

Two location-specific applications of these strategies may be cited: first, where human settlement is prohibited because of prevalence of disease; and second, where year-round water resources remain to be developed.

The land-expanding effects of disease eradication are relatively straightforward. The eradication of onchocerciasis (or river blindness) in the Volta Valley areas shared by Upper Volta,

Ghana, Togo, and Benin, has permitted farmers to settle on large tracts of previously under-utilized land. Even if the new settlers do not use improved agricultural technologies, but merely apply in the new area farming techniques they currently know, overall agricultural growth is likely to occur. That is particularly so if the settlers come from overcrowded areas where land availability constrained effective use of labor. Eliminating the tsetse fly from other savanna areas with good potential for the expansion of mixed farming holds promise of similar growth in production there--again, with only a minimum of agricultural innovation.

Irrigation efforts, on the other hand, not only expand the effective land area but also imply better exploitation of scarce water resources and more efficient use of available labor by enabling the seasonal bottlenecks and lows to be smoothed out. Given the costs of developing even fairly small irrigation systems in the savanna, however, it is unlikely that increasing use of land, water, and labor resources without changing agricultural technologies or without changing cropping patterns and choices would be profitable.

Still other means for implementing a strategy aimed at exploiting underused resources are land reform and price policies, neither of which has been widely used in West Africa to promote agricultural development. The success of land reform in accomplishing that objective depends on evidence that the land to be covered in the reform is likely to be used more intensively if it is allocated to a large number of different users rather than remaining under the control of a few. Where landholders are using this resource extensively, for example, transferring use of the land to households eager to invest their labor more intensively is likely to have a production-boosting effect. Two recent attempts to alter the land-tenure situation in Senegal and in Nigeria do not appear to have been implemented with that objective in mind, however. The law in Senegal appears to strengthen the concept that the land ultimately belongs to the government, whereas the Nigerian decree at present is largely confined to urban areas and attempts to prevent inequalities in urban land distribution. The lack of attention to the production effects implied by the increase in individual land tenure may be more serious in the long run, but no efforts to address that possibility have been noted.

The impact of normal market incentives on production, on the other hand, has been the topic of numerous recent analyses. Whereas the colonial governments, operating in the days of more limited international trade, were able to use such incentives to bring about significant increases in use of local resources, many post-colonial governments in the savanna have attempted to adjust price policies to benefit consumers or government revenues rather than to provide incentives to agricultural producers. The effects of these policies, in many cases, have been to reduce farmers' incentives to exploit available resources to the fullest where their returns were likely to be negatively affected by prices and to allocate resources instead to crops or activities relatively insulated from negative-price conditions (such as growing crops for home consumption or off-farm employment) or for which price prospects were good.

One final example of a development strategy intended to

facilitate expanded utilization of underused resources without agricultural innovation is that tried in Gambia. Development of the tourist industry there was intended to capitalize on the country's beach land and to provide jobs in the nonagricultural season for rural labor otherwise unemployed during that season. It has not been so successful as hoped, although the overall contribution of the industry to the economy has been somewhat positive (Carter 1978).

In general, strategies to increase agricultural production simply by facilitating the use of currently unused or underused resources without complementary changes in agricultural technologies will have limited or very location-specific effectiveness in savanna agriculture. The scope for increasing labor or land intensity without the use of new resources from outside the farm--information, inputs, credit, tools--is determined, as we have repeatedly implied in our analyses, by the extent to which the exogenous factors affect the productivity in a given farming system.

DIRECT STRATEGIES: TARGETING NEW RESOURCES

Recognizing the limitations of more indirect strategies for promoting agricultural development, many national governments and donor agencies have developed strategies that involve the concept of targeting resources to particular problems, thereby hoping to achieve quantifiable, assured results. These strategies imply significantly greater analysis of the constraints and more nearly precise delineation of objectives. They also assume that resources can be effectively directed to alleviate the constraints, so will achieve the selected objectives in the most efficient way. Such assumptions are rarely completely fulfilled in practice, but the additional planning and political mobilization normally involved in mounting such directed efforts reinforce the political will of the governments concerned to provide the resources to the target groups.

The farming systems approach to research can be, we believe, particularly useful both in informing the planning process for such projects and in participating in efforts designed to improve the technologies used by small-scale farmers. Such efforts often imply further, developing or adapting technology, instituting more effective support or resource-delivery systems, and changing agricultural policy or program; and because they do, a farming systems approach to research can be an integral part of the process of agricultural development.

As we have shown, considerable heterogeneity exists in the farming systems of the savanna; there are, therefore, strong arguments for incorporating an FSR element into an appropriate institution. Where technology development is an important element of the agricultural development strategy, such institutionalization is, in our view, essential. Because the potential scope for activities under the rubric of farming systems research is so broad and there are so many interdependencies among the various stages of FSR, organizational linkages are significant in determining the success of that approach. Although that is an implementation issue, it also has important connotations for methodology and the definitions of farm-level constraints to development.

Farming Systems Research

Most FSR programs currently are administered through agricultural research organizations that are committed to increasing production by developing improved technologies. Thus, the focus of FSR efforts has been largely on crop research and on tailoring component research findings to farmers' situations or to determining priorities for further scientific research. Considerable experience has been gained, and guidance for designing and carrying out such programs is available.[1] Although most programs follow a roughly similar four-stage sequence of activities--descriptive/diagnostic, design, testing, and extension--many specific methods have been devised for reducing costs, time requirements, and treatment variables as well as methods for increasing accuracy, replicability, and credibility of results.

Three important principles are emerging in designing cost- and time-efficient methodologies:

1. Reducing time required to move through the four research stages. The methodologies applied, in addition to ensuring a fast turnaround, need to be practical, replicable, and inexpensive (Byerlee et al. 1981). Complex procedures that require scarce, highly qualified individuals to collect and analyze data and to design and test solutions need to be avoided as much as possible (Zandstra 1979a). There are, however, limits to reducing the length of time required to obtain results, particularly if the body of knowledge is weak.
2. Maximizing the return from such research by making results more widely applicable. This means defining recommendation domains in terms as broad as possible. The extent to which improved systems can be transferred or extrapolated to other areas directly affects efficiency. Sequential design systems should be used.
3. Using "second best" or "best of readily available solutions." Traditionally, research in agriculture has emphasized the concept of developing optimal practices. When one considers the heterogeneity existing in the "total" environment, however, costs in terms of finance and time to obtain optimal recommendations for each possible variation--tailoring to individual farming households, for example--would be astronomical. Increasingly, therefore, the emphasis of FSR is on developing improved technologies that are better than most, but not necessarily the best, for each environment. In other words, the process is "nonperfectabilitarian" and when used optimal improved practices are not envisioned (Winkelmann and Moscardi 1979).

Some of the ways these principles are being applied in each phase of the research sequence illustrate the options that might be considered.

The descriptive and diagnostic stage. Initially, decisions have to be made concerning the geographical target area(s) on which to focus the FSR program. Criteria used often include selecting an area that: first, fits within the framework of governmental needs and priorities; second, will be conducive to extending the results over a broad area to enhance the multiplier effects of the research; and third, is an area in which credible results can be achieved within a reasonable time. Normally, that implies a less than national focus at the outset. In delineating boundaries of the target area, the researcher frequently uses criteria based on differences in administrative areas or agroecology or some compromise between the two. Access to urban markets, as was used in the Zaria village studies, is also sometimes the basis for noncontiguous area selection.

Classifying farming families within the target area into homogeneous subgroups[2] or recommendation domains involves at least two and usually three steps. Careful examination of secondary data available on the area, followed by reconnaissance (Hildebrand 1979b) or exploratory surveys (Collinson 1979), has been a useful and fairly timely approach to that task. The surveys are often informal and consist of short field tours or sondeos (Hildebrand 1979b). Multidisciplinary teams working in an interdisciplinary framework travel throughout the target area talking with representatives of policy-making organizations, farmer-contact agencies, community leaders, and farm families. Such discussions help in delineating relevant subgroups of farming families, in analyzing current farming systems, and in postulating possible types of developmental strategies potentially useful to farming families and consistent with their goals. In such exploratory surveys, interaction is required not only with people in the target area but also among members of the FSR team. The efficiency with which these reconnaissance surveys can be carried out--six to ten days in the case of the sondeo (Hildebrand 1979b)--is largely a function of the experience of the team in FSR and their familiarity with the target area.

A third step toward farm-household classification often involves a more formal, structured survey, administered by enumerators to the target population to verify tentative insights gained in the exploratory survey. Design of formal surveys involves making trade-offs between cost and time efficiency, on one hand, and accuracy, on the other. For those concerned about greater efficiency, the formal survey consists of a single interview with each participant in a representative sample of farmers. Such an approach is particularly justified when both the area and the number of variables to be considered are limited. Emphasis on accuracy or on depth of understanding, in contrast, calls for frequent interviews over a long time, usually one year, particularly for data that are continuous and nonregistered (such as labor flows or crop disposal patterns), in contrast with those that are single-point and registered in nature (such as purchase of fertilizer) (Collinson 1972; Lipton and Moore 1972).

Single-visit interviews of a large number of farmers are increasingly being undertaken to minimize sampling errors in the first instance. Complementing such surveys with in-depth and frequent interviewing of a limited number of farmers can then minimize measurement errors. The frequent-interviewing approach

(Hart 1979) usually is carried on concurrently with later stages of the FSR program. Particular emphasis is usually placed on including farming families who participate in the testing stage of FSR. That combination of single and frequent interviews has the advantage of minimizing delays in moving from the descriptive to the design and testing stages. In addition, accurate, quantitative information collected during the testing stage is particularly useful in comparing the existing system with the improved system.

The design stage. Whichever diagnostic method is selected, preliminary ideas on the priorities for research are expected to evolve from the descriptive and diagnostic stage. Collinson (1981) has suggested the following procedures for designing improved practices on the basis of diagnostic information:

1. The experimental variables should involve practices in which farmers' management is flexible and those in which ex ante evaluation suggests room for investigating what is available in the body of knowledge. Flexibility in management is improved when there are underused resources, whereas the potential for increased productivity of resources is particularly important for those that are most limiting.
2. The feasible range of treatments for such variables is indicated by the flexibility that exists. Some flexibility could be introduced, for example, by assuming that institutional support systems could change--that is, be a variable rather than a parameter. An institutional source of credit, for example, could be made temporarily available to supplement the cash flow of the farm business, if it is expected that such facilities will be made available in the future. In developing improved practices, researchers usually should consider the existing or definitely expected infrastructural support system, unless (as we discuss later) the potential exists for the FSR team to influence the support systems.
3. The parameters in the experimental process should be those not potentially subject to manipulation and should be as representative as possible of practical farming conditions.

The design stage is usually implemented at experiment stations, particularly if agronomic variables are to be emphasized in treatments. There may however be an overlap between the design and testing stages of FSR. In fact some design work can, and does, take place in researcher-managed trials at the farm level. Actual experimentation at the design stage can be reduced substantially or even eliminated if the body of knowledge is well developed and ex ante evaluation of the technology packages increases expectations of fairly robust results.

Testing. The testing phase of an FSR program is often the beginning of farmers' active involvement. As we emphasized at some length in Chapter 7, the farmer-managed trials are a key stage in developing improved agricultural technologies suitable for the totality of a farming system--water sources as well as pest

240

complexes, labor availability as well as equipment.

The line between researcher-managed and farmer-managed trials is sometimes blurred in practice, particularly where the technology being tested is totally unfamiliar to the farmers and information transfer is critical to their participation. Often, researchers' well-meaning intentions to ensure a successful production outcome or to use the test as a demonstration device lead to an intrusive role in the farmer-managed field trials. But it should be recognized that such intrusion may bias the results of the tests and reduce the potential for replicability.

Extension. In some ways extension has been the phase of FSR activity least well explored. The assumption often has been that successful farmers' test results will be automatically fed into the regular extension services present in most developing countries. Where extension agents have been involved in the FSR program all along, the chances of that happening are greater than where there has been little or no involvement. But it is in this area that organizational linkages of the FSR program are crucial and in which the credibility of the FSR program will be made or broken.

One direct way to examine the factors necessary for disseminating technologies developed and tested through the FSR program is through methods such as the guided change project discussed in Chapter 7. In that project, the feasibility of farmers' acceptance of technology packages with or without credit and with or without extension services was tested by establishing village-level programs in a dozen villages. Extension information was found to have a positive effect on farmers' willingness to use both credit and improved technology packages that included fertilizer, seeds, and seed dressings.

Credibility and efficiency. Monitoring evaluation efforts is another way to get at a basic issue of most FSR programs: how to establish credibility. The FSR approach in the developing world has been gathering momentum only since the 1970s; credibility problems therefore remain in both professional[3] and practical senses.

It would seem that the most logical way to compare the relative merits of FSR programs and research programs of a more conventional nature would be to look at cost in relation to returns. This is, of course, an empirical question. Although we hypothesize that FSAR will have a higher benefit-cost ratio in raising small farmers' productivity than commodity and disciplinary approaches will, only monitoring the use of technologies developed and extended as a result of a FSR program can provide needed data. We suggest that that is possible, although we are not sure of the relevance of the question. Maximum effectiveness of both types of research is achieved when they are undertaken together. An FSR program cannot exist without continued scientific attention to commodity or disciplinary (soils, hydrology, etc.) research questions as well.[4]

Nevertheless, in estimating the returns from FSR, the obvious criterion is measurement of the improvement in the welfare of farming families. Measuring rural welfare, however, is very difficult. For example, FSR may directly or indirectly increase the welfare of farming families—indirectly by reorienting research

priorities of other research programs so they later contribute to increasing farmers' welfare. Unfortunately, the potential of such feedback is often ignored in evaluating FSR contributions, possibly because it is difficult, if not impossible, to quantify in the short run.
Despite considerable potential returns, efforts to reduce the time and costs of producing credible FSR results are necessary if this approach is ever to be applied to the actual benefit of a significant portion of the farm population in the developing world. Unlike the results of the Green Revolution, the results of FSR are likely to be less spectacular because of the step-by-step modification rather than the sudden transformation of farming systems.[5] As a result, the credibility FSR achieves is likely to be heavily influenced by how efficiently research funds are used. And this again raises the issues of the organization, location, and methodological choices of a particular FSR program. Even where there is a commitment at top governmental levels, those administering FSR programs in agricultural research institutions may be frustrated by the nonresponsiveness of government bureaucracies accustomed to looking to the central ministry headquarters as the source of all wisdom and direction. The organization of the agricultural development effort may already be so fragmented along regional, commodity, discipline, and functional lines that opposition to initiating new FSR programs--to say nothing of the reluctance to implement the results of existing FSR work in a particular area--may be great.
Further, very few parts of the developing world are unscarred by development projects. Those that have failed often leave a residue of bitterness and opposition among the local residents to all things connected with the government. When going into areas where there are on-going projects that are having difficulties, or which are operating on completely different principles, FSR teams are faced with the worst of both worlds: the opposition of the local people and suspicions of the implementing agencies that do not wish to be discredited. Yet on-going projects often provide an opportunity for FSR to make a contribution to farming families' welfare by modifying practices that are being recommended or by providing the evidence needed to terminate the project.

The Farming Systems Perspective

If there is political, and thus budgetary, commitment to implementing the farming systems approach to research but institutional arrangements are such that the establishment of a formal farming systems research program in an existing agricultural research organization does not make sense, other institutional linkages may be more appropriate.
Physically locating a team in an agricultural planning unit, for example, rather than in a crop-research organization, might increase the potential for a farming-system perspective at the policy level to assist in the design of strategies to achieve desired agricultural growth objectives, particularly when they are focussed on improving the welfare of farming households. Such a location is also likely to lead to an emphasis on variables other

than those related to improving agricultural technologies, such as the links between policies and support institutions external to the farming households; that is, the set of exogenous variables. Experience with this type of organizational location is, however, comparatively rare. One notable exception is the Caqueza project in Colombia: an FSR group worked with credit institutions serving the project area in designing schemes to deal with risk aversion (Zandstra et al. 1979).

The widely endorsed objective of ensuring equitable growth is usefully considered from a farming systems perspective. As we have noted, there is evidence that, given agricultural technologies currently available in the savanna and the current distribution of resources, inequality is both present and growing. The guided change research project discussed in Chapter 7, however, provides persuasive evidence that attempts to ensure equitable access to inputs and support systems can result in greater production--and greater village income--than would be expected if no efforts were taken to provide such access.

A national development objective also can be forwarded by calling on farming-systems-research teams at international institutions. In India, farming systems researchers from the International Cereals Research Institute for the Semi-Arid Tropics (ICRISAT) involved local bankers in the testing stage of their program with a view to obtaining their assessments of the feasibility of soil and water management technology and, in particular, of the prospects for loans to finance items such as the tropiculteur--an animal-drawn implement. As a result of this inter-organizational collaboration, this implement is now an approved item for credit in the Indian banking system.

Through such linkages, FSR teams can consider improvements that officially may be outside the mandate of a technical research institute and also may play an interventionist role in influencing the support systems serving agriculture. This role goes beyond the somewhat submissive FSR approach in which the support systems are accepted as parameters and improved technologies are selected and tested on the basis of assumptions about levels of support that are expected to be provided.

Where such an interventionist approach is possible, detailed information generated through FSAR could be important for identifying changes in policies that would complement the introduction of improved practices. It should be noted, however, that experience to date has been concentrated on technology development. There are few guidelines for setting up micro-oriented farming-systems teams to bring this perspective to organizations having broader focus. But it would certainly be worth trying.

Equity considerations also play a role in encouraging the development of scale-neutral agricultural technology. Our discussion of ox-plowing as a new technology appropriate to crop production conditions in some parts of the savanna, however, provides ample evidence that many practical agricultural innovations cannot be used profitably by all farmers, regardless of land-holding size. Similarly, a motorized cotton sprayer may not be a suitable technology for farming households whose need to ensure food supplies causes them to plant so late that yields will be substantially reduced anyway. Other than using scale neutrality as the single appropriate test for an equitable technology, it is up to the researchers concerned with increased agricultural output to consider two aspects of the farming systems that are expected to be "customers" for the technology: first, which farming systems will the technology fit and second, are the support systems in place which will ensure equal access to all the potential farming systems?

From the view of a policy-maker concerned with equitable growth, maximum rural welfare, and reduced chances of exacerbating current inequalities, the answer to the first question must, in our view, be "all" or at least "a majority." The second answer must be "yes."

Interestingly, many advocates of agricultural development geared to equitable growth have not asked those questions. Thus, technologies recommended for extension or replication have not fit farming systems. The temporary, often expensive, support systems set up for the duration of the project often prove to be unsustainable in the long run. The World Bank's recent review of African development has recognized, as we have noted, the futility of expecting agricultural growth or improved welfare in such cases, no matter how many resources are targeted to accomplish such objectives (World Bank 1981a). The argument has now been made that project implementation should be concentrated in the more promising areas pending development of relevant improved technologies for the more poorly endowed areas. Although seemingly inequitable, it may be a realistic assessment because the economies of the countries in the region are such that equitability cannot be achieved through welfare redistribution programs but will have to be based on income-generating opportunities for the more disadvantaged. Even when decisions have been made on inter-areal allocation, difficult decisions remain to be made on allocating the developmental resources within the area--for example, should they be concentrated or dispersed?

But simply ignoring those concerns and concentrating on short-term exploitation of comparative advantage might have the effect of exacerbating already existing inequalities in rural welfare. In the West African savanna, it would mean regional concentration in the areas of higher rainfall, accepting the instability of return inherent in the reliance on export markets, and a growing dependence in many places on imported food. Further, if no attempts were to be made within these areas of comparative advantage to ensure equitable access to resources (such as credit, improved seeds, and fertilizers) and present policies on subsidies on inputs were maintained, there would be some reason to believe that both overall production and welfare would be less than would be possible if greater efforts were made to ensure such access, particularly to those households with underused resources. Determining who these households are and verifying that improved technologies will fit these systems are the essence of the farming systems approach to research.

IMPLICATIONS FOR POLICY IN NIGERIA

The savanna areas in Nigeria continue to be important for food production and employment, even though petroleum exports have displaced groundnuts and cotton in the national income picture. Over 30 percent of the calories in the average national diet were estimated to have been produced in the savanna (World Bank 1979). Some 50 million people were likely to have depended upon food grain production from their own and other small farms in 1980. If one considers the employment in agro-allied industries in which domestic inputs are used, significant numbers of urban Nigerians also depend on the continued functioning of these farming systems for their jobs.

Idachaba (1980) points out that Nigeria has invested less than 1 percent of its total public sector expenditure over the last ten years in its agricultural research establishment, and a sharply declining amount of the total allocation to agriculture has been directed to research. Though that probably represents a relatively greater public sector rate of allocation to agriculture research than has been made by many of the other savanna states, it has not been adequate to achieve an annual rate of growth in agricultural productivity that exceeds the estimated rate of population growth.[6] Research expenditures on commodities important to the rainfed savanna farmers have been roughly 10 to 15 percent of the total federal allocations to research; the area devoted to these commodities is estimated to have been approximately 70 to 85 percent of the total food crop area in the 1970s (World Bank 1979).

Yet Nigeria has expressed commitment to the farming systems approach to research. As we noted in Chapter 1, the government-supported Institute for Agricultural Research has supported FSAR-type efforts with regular budget allocations since the early 1970s. A number of Nigerian professionals were trained in multidisciplinary approaches to research on improved agricultural technologies and have continued working in this mode as they have taken over the reins of program administration as well as of research. The efforts to develop and use a farming systems approach to research discussed in this book were also followed by programs at other Nigerian universities such as Nsukka and Ife, and at the International Institute for Tropical Agriculture (IITA) located at Ibadan. The methods used at each institution have been somewhat different, but the principles of the FSAR have been shared.

Nigeria is now in the enviable position of being able to expand and to strengthen its commitment to this approach--having financial, institutional, and professional resources to draw upon. Whether it chooses to ensure that policies are in place to facilitate the greater use of its underexploited resources is, of course, a decision we can neither influence nor predict. Nevertheless, based on discussion and analysis in this book, we find it imperative that policies be developed to permit the more efficient use of resources in order to increase productivity of the agricultural sector and thus to increase output, incomes, and welfare within this sector. Based upon the farming systems approach to research, key ingredients of these policies should be broad participation by all segments of the rural sector, expectations geared to achieving modest gains rather than breakthroughs, and attention to the needs for risk-aversion and security of farming households. From a farming systems perspective, these policies are likely to facilitate rather than to dictate, to provide for incremental gains rather than revolutionary changes, and to be realistic rather than unattainably visionary.

NOTES

1. See, for example, Asian Cropping Systems Working Group (1979); Byerlee and Collinson (1980); Flinn (1979); Gilbert, Norman, and Winch (1980); Harrington (1982); Perrin et al. (1976); Shaner, Philipp, and Schmehl (1981); Technical Advisory Committee (1978); and Zandstra et al. (1981).

2. That is, homogeneous with respect to a selected set of variables, not with regard to all possible characteristics.

3. For example, often "good" agronomic research is that which produces a low coefficient of variation. An agronomist setting up a program of field trials would, therefore, tend to favor fewer trials and more replications per trial. An economist, on the other hand, to achieve results representative over a wider area, would tend to favor more trials and few replications--given limited research resources (Crawford 1980).

4. Although conceptually there is a complementary relationship between FSR and experiment station-based research, the relationship sometimes appears to be competitive because of limited research funds available for developing improved agricultural technology.

5. In aggregate the benefits of FSR may be significant because large numbers of farming families adopt the changes.

6. For the food deficit to the eliminated by 1985, domestic production of food crops would need to increase at an annual rate of 6.6 percent between 1980 and 1985 and fisheries and livestock at 11.3 percent annually during the same period (Gusau 1981).

References

Abalu, G. O. I. The Food Situation in Nigeria: An Economic Analysis of Sorghum and Millet. Samaru Miscellaneous Paper, no. 80. Zaria, Nigeria: Institute for Agricultural Research, Ahmadu Bello University, 1978.

Abalu,, G. O. I., and D'Silva, B. C. "Nigeria's Food Situation: Problems and Prospects." Food Policy 5 (1980):49-60.

Abalu, G. O. I., and Harkness, C. "Traditional Versus Improved Groundnut Production Practices: An Economic Analysis of Groundnut at the Farmers' Level in the North of Nigeria." Zaria, Nigeria: Institute for Agricultural Research, Ahmadu Bello University, 1976. (Mimeographed.)

Abalu, G. O. I., and Ogungbile, A. O. "Land Tenure, Land Resource Use and Agricultural Development in Nigeria." Paper presented at the Twelfth Annual Conference of the Agricultural Society of Nigeria, Ile-Ife, Nigeria, 5-10 July, 1976.

Abbott, J. C. "The Efficiency of Marketing Board Operations." In The Marketing Board System, pp. 231-240. Edited by H. M. A. Onitiri, and D. Olatunbosun. Ibadan, Nigeria: Nigeria Institute for Social and Economic Research, 1974.

Adamu, S. O. "The Stabilization Policy of the Marketing Boards in Nigeria, 1948-62." Nigerian Journal of Economics and Social Studies 12 (1970):322-340.

Agbonifo, P. O. "Agro-Industrialism in the Greater Zaria Area: The Introduction of Dry Season Tomato Production along the Kubanni and Shika River Valleys." M.Sc. thesis, Ahmadu Bello University, Nigeria, 1974.

Agbonifo, P. O., and Cohen, R. "The Peasant Connection: A Case Study of the Bureaucracy of Agri-Industry." Human Organization 35 (1976):367-379.

Agricultural Planning Division. "Nigerian Agricultural Input Survey, 1970-73." Statistical Series FDA/PLD, no. 8. Lagos, Nigeria: Agricultural Planning Division, Federal Department of Agriculture, 1974.

Akintomade, M. A. "A Comparative Analysis of the Marketing Board System and Other Arrangements for Commodity Marketing." In The Marketing Board System, pp. 88-134. Edited by H. M. A. Onitiri, and D. Olatunbosun. Ibadan, Nigeria: Nigerian Institute for Social and Economic Research, 1974.

Alkali, M. "Mixed Farming: Need and Potential." In Livestock

247

Development in the Dry and Intermediate Savanna Zones, pp. 36-41. Edited by Institute for Agricultural Research. Zaria, Nigeria: Ahmadu Bello University, 1969.

Ancey, G. Facteurs et Systemès de Production dans la Sociéte Mossi d'Aujourd'hui: Migrations, Travail, Terre et Capital. Ouagadougou, Upper Volta: ORSTOM, 1974.

Ancey, G. Niveaux de Décision et Fonctions Objectif en Milieu Africain. AMIRA, no. 3. Paris, France: INSEE, 1975.

Andrews, D. J. "Intercropping with Sorghum in Nigeria." Experimental Agriculture 8 (1972):139-150.

Andrews, D. J. "Responses of Sorghum Varieties to Intercropping." Experimental Agriculture 10 (1974):57-63.

Anthonio, Q. B. O. "The Marketing of Staple Foodstuffs in Nigeria: A Study in Pricing Efficiency." Ph.D. dissertation, London University, England, 1968.

Apeldoorn, G. J. van. Perspectives on Drought and Famine in Nigeria. Hemel Hempstead, England: George Allen and Unwin, 1961.

Asian Cropping Systems Working Group. Network Methodology and Cropping System Research in Indonesia. Bogor, Indonesia: Central Institute for Agriculture, 1979.

Awolola, M. D., and Buntjer, B. J. "The Introduction of New Crop Growing Technology: Opinions and Reactions." Samaru Agricultural Newsletter 18 (1976):123-130.

Baier, S., and King, D. J. "The Development of Sahelian Economics: A Case Study of Hausa-Touareg Interdependence." Land Tenure Center, University of Wisconsin, Newsletter 45 (1974):11-22.

Baker, E. F. I. "Research on Mixed Cropping with Cereals in Nigerian Farming Systems: A System for Improvement." In Proceedings, International Workshop on Farming Systems, 18-21 November, 1974, Hyderabad, pp. 287-302. Edited by International Crop Research Institute for the Semi-Arid Tropics. Hyderabad, India: ICRISAT, 1975.

Baker, E. F. I. "Mixed Cropping in Northern Nigeria. 3. Mixtures of Cereals." Experimental Agriculture 15 (1979):41-48.

Baker, E. F. I., and Yusuf, Y. "Mixed Cropping Research in Northern Nigeria." In Intercropping in the Semi-Arid Areas, 17-18. Edited by J. H. Monyo, A. D. R. Ker, and M. Campbell. Ottawa: International Development Research Center, 1976.

Balcet, J. C., and Candler, W. "Farm Technology Adoption in Northern Nigeria." 2 vols. Washington: Cornell University and World Bank, 1981. (Mimeographed.)

Bank Manager. Koutiala, Mali. Interview, September 1978.

Barrett, V.; Lassiter, G.; Mayabouti, D.; and Stickley, T. Animal Traction Credit in Six Intensive Zones of the Eastern ORD of Upper Volta. Report, no. 2. Fada N'Gourma, Upper Volta: Eastern Regional Development Organism (ORD-EST), 1978.

Beals, R. E., and Menzies, C. F. "Migrant Labor and Agricultural Output in Ghana." Oxford Economic Papers 21 (1970):109-127.

Beeden, P.; Hayward, J. A.; and Norman, D. W. "A Comparative Evaluation of Ultra Low Volume Insecticide Application of Cotton Farms in the North Central State of Nigeria." Nigerian Journal of Crop Protection 2 (1976):23-29.

Beeden, P.; Norman, D. W.; Pryor, D. H.; Kroeker, W. J.; Hays, H.

M.; and Huizinga, B. The Feasibility of Improved Sole Crop Cotton Technology for the Small-Scale Farmer in the Northern Guinea Savanna of Nigeria. Samaru Miscellaneous Paper, no. 61. Zaria, Nigeria: Institute for Agricultural Research, Ahmadu Bello University, 1976.

Beets, W. C. "Understanding the Farmer Before Introducing New Cropping Systems." World Crops 31 (1979):136-141.

Belloncle, G. Le Crédit Agricole dans les Pays d'Afrique d'Expression Française au Sud du Sahara. Rome, Italy: FAO, 1968.

Benneh, G. "Population, Food and Nutrition in a Northern Savannah Village of Ghana." Food and Nutrition in Africa 12 (1973):34-47.

Bernus, E. "L'Evolution Récente des Relations entre Eleveurs et Agriculteurs en Afrique Tropicale: L'Example du Sahel Nigérien." Cahiers ORSTOM 11 (1974):137-143.

Binswanger, H. P., and Ruttan, V. W. Induced Innovation: Technology, Institutions and Development. Baltimore: Johns Hopkins University Press, 1978.

Binswanger, H. P.; Jodha, N. S.; and Barah, B. C. "The Nature and Significance of Risk in the Semi-Arid Tropics." In Socioeconomic Constraints Development of Semi-Arid Tropics, pp. 303-316. Edited by J. G. Ryan, and H. L. Thompson. Hyderabad, India: International Crops Research Institute for the Semi-Arid Tropics, 1980.

Black, C. C. "Ecological Implications of Dividing Plants into Groups with Distinct Photosynthesis Production Capacities." Advanced Ecological Research 7 (1971):87-114.

Bouman, F. J. A. "Indigenous Savings and Credit Societies in the Third World: A Message." Savings and Development 4 (1977): 181-219.

Bungudu, L. M. "The Storage of Farm Products by Farmers in My Village." Samaru Agricultural Newsletter 12 (1970):2-10.

Bunting, A. H. Science and Technology for Human Needs, Rural Development and the Relief of Poverty. IADS Occasional Paper. New York: International Agricultural Development Service, 1979.

Buntjer, B. J. "The Changing Structure of Gandu." In Zaria and Its Region: A West African Savannah City and Its Environs, Department of Geography Occasional Paper, no. 4, pp. 157-169. Edited by M. J. Mortimore. Zaria, Nigeria: Department of Geography, Ahmadu Bello University, 1970a.

Buntjer, B. J. "The Dissemination of New Cotton Prices in Some Agricultural Communities in Zaria Province." The Nigerian Agricultural Journal 7 (1970b):81-90.

Buntjer, B. J. "Aspects of the Hausa System of Cultivation around Zaria." Samaru Agricultural Newsletter 13 (1971):18-20.

Buntjer, B. J. "Whom to Blame, the Farmer or the Extension Worker?" Samaru Agricultural Newsletter 14 (1972):8-13.

Byerlee, D. K., and Collinson, M. P. Planning Technologies Appropriate to Farmers: Concepts and Procedures. Londres, Mexico: Economics Program, Centro Internacional de Mejoramiento de Maiz y Trigo, 1980.

Byerlee, D.; Biggs, S.; Collinson, M.; Harrington, L.; Martinez, J.

C.; Moscardi, E.; And Winkelmann, D. "On-Farm Research to Develop Technologies Appropriate to Farmers." In The Rural Challenge, IAAE Occasional Paper, no. 2, pp. 170-180. Edited by M. A. Bellamy, and B. L. Greenshields. Aldershot, England: Gower, 1981.

Caldwell, J. C. "Determinants of Rural-Urban Migration in Ghana." Population Studies 22 (1968):361-377.

Campbell, D. J. "Strategies for Coping with Drought in the Sahel: A Study of Recent Population Movements in the Department of Maradi, Niger." Ph.D. thesis, Clark University, Canada, 1977.

Carter, J. "The Socio-Economic and Cultural Impact of Tourism upon the Gambia." Final draft report for the World Bank. Washington, World Bank, 1978. (Mimeographed.)

Castle, E. M. "A Framework for Rural Development." Culture and Agriculture 3 (1977):1-6.

Caswell, G. H. "The Storage of Cowpea in the Northern States of Nigeria." Proceedings of the Agricultural Society of Nigeria 5 (1968):4-6.

Central Bank of Nigeria. "Food Supply in Nigeria, 1960-1975." Economic and Financial Review 15 (1977):12-19.

Central Bank of Nigeria. Nigeria's Principal Economic and Financial Indicators 1970-1977. Lagos, Nigeria: Central Bank of Nigeria, 1978.

Central Bank of Nigeria. Annual Report and Statement of Accounts. Lagos, Nigeria: Central Bank of Nigeria, various issues.

Chambers, R.; Longhurst, R.; Bradley, R.; and Feachem, R. Seasonal Dimensions to Rural Poverty: Analysis and Practical Implications. Discussion Paper, no. 142. Brighton, England: Institute of Development Studies, University of Sussex, 1979.

Chaminade, R. "Recherches sur la Fertilité et la Fertilisation des Sols en Régions Tropicales." Agronomie Tropicales 27 (1972):891-904.

Charlick, R. B. "Power and Participation in the Modernization of Rural Hausa Communities." Ph.D. thesis, University of California, 1974.

Charreau, C. "Soils of Tropical Dry and Dry-Wet Climatic Areas and Their Use and Management." Ithaca: Cornell University, 1974a. (Mimeographed.)

Charreau, C. "Systems of Cropping in the Dry Tropical Zone of West Africa with Special Reference to Senegal." In International Workshop on Farming Systems in the Semi-Arid Tropics, pp. 287-302. Edited by International Crops Research Institute for the Semi-Arid Tropics. Hyderabad, India: ICRISAT, 1974b.

Charreau, C. "Some Controversial Technical Aspects of Farming Systems in Semi-Arid West Africa." Dakar, Senegal: ICRISAT Regional Office, 1978. (Mimeographed).

Charreau, C., and Fauck, R. "Mise au Point sur l'Utilisation Agricole des Sols de la Region de Séfa (Casamance)." Agronomie Tropicale 25 (1970):151-191.

Charreau,C., and Nicou, R. "L'Amélioration du Profil Cultural dans les Sols Sableux and Sablo-Argileux de la Zone Tropical Sèche Ouest Africaine et Ses Incidences Agronomiques." Agronomie Tropicale 26 (1971):209-255, 565-631, 903-978, 1183-1237.

Cleave, J. H. African Farmers: Labor Use in the Development of

Smallholder Agriculture. Washington: Praeger, 1974.
Clough, P. "The Relationship between Rural Poverty and the Structure of Trade." Oxford, England: Oxford University, 1977. (Mimeographed.)
Cocheme, J., and Franquin, P. A Study of the Agroclimatology of the Semi-Arid Areas South of the Sahara in West Africa. Rome: Food and Agriculture Organization, 1967.
Collinson, M. P. Farm Management in Peasant Agriculture: A Handbook for Rural Development Planning in Africa. New York: Praeger, 1972.
Collinson, M. P. "Understanding Small Farmers." Paper given at Conference on Rapid Rural Appraisal, Institute of Development Studies, University of Sussex, England, 4-7 December, 1979.
Collinson, M. P. "Micro-level Accomplishments and Challenges for the Less Developed World." In Rural Change: The Challenge for Agricultural Economists, pp. 43-53. Edited by G. Johnson and A. Maunder. Aldershot, England: Gower, 1981.
Connell, J., and Lipton, M. Assessing Village Labor Situations in Developing Countries. Delhi, India: Oxford University Press, 1977.
Crawford, E. Department of Agricultural Economics, Michigan State University, East Lansing. Letter, January 1980.
Critchfield, R. Freelance Writer. Letter to L. Hardin, Ford Foundation, 15 February, 1979.
Crook, J. M., and Ward, M. "The Quelea Problem in Africa." In The Problems of Birds as Pests, pp. 211-229. Edited by R. K. Murton, and K. N. Wright. London, England: Academic Press, 1968.
Czarnocki, K. "The Relationship between Crops and Livestock in the Sudanese Zone of West Africa." African Bulletin 19 (1973):115-122.
Damachi, V. G., and Siebel, H. D., eds. Social Change and Economic Development in Nigeria. New York: Praeger, 1973.
Delgado, C. L. Livestock versus Foodgrain Production in Southern Upper Volta: A Resource Allocation Analysis. Ann Arbor: Center for Research on Economic Development, University of Michigan, 1978.
Diarra, S. "Les Problèmes de Contact entre les Pasteurs Peuls et les Agriculteurs dans le Niger Central." In Pastoralism in Tropical Africa, pp. 284-297. Edited by T. Monod. Oxford, England: International African Institute, Oxford University Press, 1975.
Dillon, J. L., and Hardaker, J. B. Farm Management Research for Small Farmer Development. FAO Agricultural Services Bulletin, no. 41. Rome, Italy: Food and Agriculture Organization, 1980.
D'Silva, B. C., and Raza, M. R. ."Integrated Rural Development in Nigeria - The Funtua Project." Food Policy 5 (1980):282-297.
D'Silva, B. C.; Raza, M. R.; Ejiga, N. O. O.; and Bogunjoko, I. O. "Socio-Economic Impact of the Funtua Agricultural Development Project." Zaria, Nigeria: Department of Agricultural Economics and Rural Sociology, Ahmadu Bello University, 1980. (Mimeographed.)
Dubois, J. P. "Les Sérer et la Question des Terres Neuves au Sénégal." Cahiers ORSTOM (Série Sciences Humaines) 12

(1975):81-120.

Dunsmore, J. R.; Rains, A. B.; Lowe, G. D. N.; Moffatt, D. J.; Anderson, I. P.; and Williams, J. B. The Agricultural Development of the Gambia: An Agricultural, Environmental and Socio-Economic Analysis. Land Resource Study, no. 22. Tolworth, England: Ministry of Overseas Development, Land Resource Development Center, 1976.

Echard, N. "Socio-Economic Study of the Ader Doutchi Majya Valleys." Etudes Nigériennes, no. 15. Niamey, Niger: IFAN, 1964.

Edache, O. A. "Motivational Factors Related to the Adoption of Improved Farm Practices: A Study of the Sorghum Program of the Kano State Pilot Phase of the National Accelerated Food Production Program." M.S. thesis, Kansas State University, 1978.

Ejiga, N. O. O. "Economic Analysis of Storage, Distribution and Consumption of Cowpeas in Northern Nigeria." Ph.D. dissertation, Cornell University, 1977.

Elias, T. Nigerian Land Law. London, England: Routledge and Kegan Paul, 1972.

Ernst, K. Tradition and Progress in the African Village: The Non-Capitalist Transformation of Rural Communities in Mali. London, England: C. Hurst, 1976.

Essang, S. M. "Achieving Food and Population Balance in Nigeria." In Food and Population: Priorities in Decision Making, pp. 103-113. Edited by T. Dams, K. E. Hunt, and G. J. Tyler. Farnborough: Saxon House, 1978.

Falusi, A. O. "Economics of Fertilizer Distribution and Use in Nigeria." Ph.D. dissertation, Cornell University, 1973.

Falusi, A. O., and Williams, L. B. Nigeria Fertilizer Sector: Present Situation and Future Prospects. Technical Bulletin, no. T-18. Muscle Shoals: International Fertilizer Development Center, 1981.

FAO. Calorie Requirements: Report of the Second Committee on Calorie Requirements. Nutritional Studies, no. 15, Rome, Italy: Food and Agriculture Organization, 1957.

FAO. Report to the Government of Nigeria on Seed Production and Distribution. Rome, Italy: Food and Agriculture Organization, 1970.

Fauck, R.; Moureaux, C.; and Thomann, C. "Bilans de l'Evolution des Sols de Séfa (Casamance, Sénégal) après Quinze Années de Culture Continue." Agronomie Tropicole 24 (1969):263-301.

Faulkingham, R. H. "Ecological Constraints and Subsistence Strategies. The Impact of Drought in a Hausa Village: A Case Study from Niger." In Drought in Africa 2, African Environment Special Report no. 6, pp.? Edited by D. Dalby, R. J. Harrison Church, and F. Bezzaz. London, England: International African Institute, Oxford University Press, 1977.

Faye, J. Problématique d'un Thème Technique Agricole: Le Labour de Fin de Cycle avec Enfouissement des Pailles. Bambey, Sénégal: CNRA/ISRA, 1977.

Faye, J. Experimental Units, Kaolack, Senegal. Interview, September 1978.

Faye, J., and Niang, M. "An Experiment in Agrarian Reconstruction and Senegalese Rural Space Planning." African Environment 2 (1977):143-153.

Federal Office of Statistics. Digest of Statistics. Lagos, Nigeria: Federal Office of Statistics, various issues.

Federal Republic of Nigeria. Third National Development Plan 1975-80. Vol. 1. Lagos, Nigeria: The Central Planning Office, Federal Ministry of Economic Development, 1975.

Flinn, J. C. "Focusing Farming Systems Research on Smallholder Agriculture: Experiences from West Africa." In The Adaptation of Traditional Systems of Agriculture: Socio-Economic Problems of Urbanisation. Development Studies Centre Monograph, no. 11, pp. 257-276. Edited by Fisk, E. K. Canberra, Australia: Australian National University, 1978.

Flinn, J. C. "Agroeconomic Considerations in Cassava Intercropping Research." In Intercropping with Cassava: Proceedings of an International Workshop Held at Trivandrum, India, 27 November - 1 December, 1978. Edited by E. Weber, B. Nestel, and M. Campbell. Ottawa, Canada: International Development Research Centre, 1979.

Fumagelli, A., and Waugh, R. Agricultural Research in Guatemala. Guatemala City: Instituto de Ciencia y Tecnologia Agricolas, 1977.

Gentil, D. "La Naissance d'un Système Coopératif: L'Example du Niger." Options Méditerranéennes 6 (1971a):88-95.

Gentil, D. Les Coopératives Nigériennes. Paris, France: Ecole Pratique de Hautes Etudes, 1971b.

Gilbert, E. H. "Marketing of Staple Foods in Northern Nigeria: A Study of the Staple Food Marketing Systems Serving Kano City." Ph.D. dissertation, Stanford University, 1969.

Gilbert, E. H.; Norman, D. W.; and Winch, F. E. Farming Systems Research: A Critical Appraisal. Rural Development Paper, no. 6. East Lansing: Department of Agricultural Economics, Michigan State University, 1980.

Giles, P. H. The Storage of Cereals by Farmers in Northern Nigeria. Samaru Research Bulletin, no. 42. Zaria, Nigeria: Institute for Agricultural Research, Ahmadu Bello University, 1965.

Goddard, A. D. "Population Movements in the Sokoto Close-Settled Zone, Northern Nigeria." Paper presented at the Eleventh International African Seminar, International African Institute, Dakar, Senegal, 27 March - 6 April, 1971.

Goddard, A. D. "Land Tenure, Land Holding and Agricultural Development in the Central Sokoto Close-Settled Zone." Savanna 1 (1972):29-41.

Grant, M. W. Nutrition Field Working Party: Food Consumption Data. Colonial Report no. 20032. London, England: HMSO, 1970.

Guggenheim, H. "Of Millet, Mice and Men: Traditional and Invisible Technology Solutions to Pool Harvest Losses in Mali." In World Food Pest Losses and the Environment, AAAS Selected Symposium 13, pp. 109-162. Edited by D. Pimental. Boulder: Westview Press, 1978.

Guinard, A. "Conservation and Improvement of Soil Fertility. Part 2." World Crops 19 (1967):29-31.

Guissou, J. Etude sur les Besoins des Femmes dans les Villages de l'AVV et Proposition d'un Programme d'Intervention.

Ouagadougou, Upper Volta: SAED, Ministère du Development Rural, 1977.
Gusau, A. "Nigeria's Green Revolution." Africa Report 26 (1981):19-22.
Hall, M. "A Review of Farm Management Research in East Africa." ECA/FAO Agricultural Economics Bulletin for Africa 12 (1970):11-24.
Hardin, L. S. "Discussion: Social Science and Related Implications." American Journal of Agricultural Economics 59 (1977):853-854.
Harrington, L. Methodological Issues Facing Social Scientists in On-Farm Farming Systems Research. Londres, Mexico: Economics Program, Centro Internacional De Mejoramiento de Maiz y Trigo, 1981.
Harriss, B. "Cereal Surpluses in the Sudano-Sahelian States." 2 vols. ICRISAT Economics Program Consultancy Report. Hyderabad, India: ICRISAT, 1978. (Mimeographed)
Hart, R. D. "One Farm System in Honduras: A Case Study in Farm Systems Research." Paper presented at Farming Systems Research and Development Workshop, Consortium for International Development, Fort Collins, Colorado, 2-4 August, 1979.
Harwood, R. R. Small Farm Development: Understanding and Improving Farming Systems in the Tropics. Boulder: Westview, 1979.
Harwood, R. R., and Price, E. C. "Multiple Cropping in Tropical Asia." In Multiple Cropping, ASA Special Publication, no. 27, pp. 11-40. Edited by the American Society of Agronomy. Madison: American Society of Agronomy, 1976.
Haswell, M. R. Economics of Agriculture in a Savannah Village. Colonial Research Study, no. 8. London, England: HMSO, 1953.
Haswell, M. R. The Nature of Poverty. London, England: Macmillan, 1975.
Hayami, Y., and Ruttan, V. W. Agricultural Development: An International Perspective. Baltimore: Johns Hopkins University Press, 1971.
Hays, H. M. The Marketing and Storage of Food Grains in Northern Nigeria. Samaru Miscellaneous Paper, no. 50. Zaria, Nigeria: Institute for Agricultural Research, Ahmadu Bello University, 1975a.
Hays, H. M. The Storage of Cereal Grains in Three Villages of Zaria Province, Northern Nigeria." Savanna 4 (1975b):117-123.
Hays, H. M. "Agricultural Marketing in Northern Nigeria." Savanna 5 (1976):139-148.
Hays, H. M., and McCoy, J. H. "Food Grain Marketing in Northern Nigeria: Spatial and Temporal Performance." Journal of Development Studies 14 (1978):182-192.
Hays, H. M., and Raheja, R. K. "Economics of Sole Crop Cowpea Production in Nigeria at the Farmers' Level Using Improved Practices." Experimental Agriculture 13 (1977):149-154.
Helleiner, G. K. Peasant Agriculture, Government, and Economic Growth in Nigeria. Homewood: Irwin, 1966.
Helleiner, G. K. "The Marketing Board System and Alternative Arrangements for Commodity Marketing in Nigeria." In The Marketing Board System, pp. 55-70. Edited by H. M. A. Onitiri, and D. Olatunbosun. Ibadan, Nigeria: Nigerian Institute of

Social and Economic Research, 1974.

Hildebrand, P. E. "Generating Technology for Traditional Farmers: A Multidisciplinary Methodology." Paper presented at Conference on Developing Economies in Agrarian Regions: A Search for Methodology, Rockefellor Foundation Conference Center, Bellagio, Italy, 4-6 August, 1976.

Hildebrand, P.E. "Generating Technology for Traditional Farmers: The Guatemalan Experience." Paper presented at International Congress of Plant Protection, Washington, D.C., 5-11 August, 1979a.

Hildebrand, P.E. Summary of the Sondeo Methodology Used by ICTA. Guatemala City: Instituto de Ciencia y Tecnologia Agricolas, 1979b. (Mimeographed)

Hill, P. Rural Hausa: A Village and a Setting. London, England: Cambridge University Press, 1972.

Holdcroft, L. E. The Rise and Fall of Community Development in Developing Countries, 1950-65: A Critical Analysis and an Annotated Bibliography. Rural Development Paper, no. 2. East Lansing: Department of Agricultural Economics, Michigan State University, 1978.

Hopkins, E. "Operation Groundnuts: Lessons from an Agricultural Extension Scheme." Institute of Development Studies Bulletin 3 (1974):59-66.

Hopkins, E. "Wolof Farmers in Senegal: A Study of Responses to an Agricultural Extension Scheme." Ph.D. thesis, University of Sussex, England, 1975.

Horowitz, M. M. "Ethnic Boundary Maintenance among Pastoralists and Farmers in the Western Sudan." Journal of Asian and African Studies 7 (1972):105-114.

Huizinga, B. "Some Thoughts on Information Systems for Small Farmer Development." Zaria, Nigeria: Institute for Agricultural Research, Ahmadu Bello University, 1978. (Mimeographed)

Huizinga, B. "An Experiment in Small Farmer Development Administration amongst the Hausa of Nigeria." Zaria, Nigeria: Institute for Agricultural Research, Ahmadu Bello University, n.d. (Mimeographed)

Huizinga, B.; Reawaruw, I. P.; Engelhard, R. J.; DeWit, T. J.; and Etuk, E. G. "A Technical Note on the Guided Change Project." Zaria, Nigeria: Institute for Agricultural Research, Ahmadu Bello University, 1978a. (Mimeographed)

Huizinga, B.; Reawaruw, I. P.; Engelhard, R. J.; DeWit, Th. J.; and Etuk, E. G. "The Guided Change Project: An Approach to Agricultural Development in Northern Nigeria; First Lessons and Experiences." Zaria, Nigeria: Institute for Agricultural Research, Ahmadu Bello University, 1978b. (Mimeographed)

Hunter, J. M. "Seasonal Hunger in a Part of the West African Savanna: A Survey of Body Weights in Nangodi, Northeast Ghana." In Markets and Marketing in Africa. Edited by Centre of African Studies. Edinburgh, Scotland: University of Edinburgh Press, 1966.

IAR. Shika Agricultural Research Station Biennial Report 1969-71. Zaria, Nigeria: Institute for Agricultural Research, Ahmadu Bello University, 1971.

IAR. Report to the Board of Governors for 1971. Zaria, Nigeria:

Institute for Agricultural Research, Ahmadu Bello University, 1972.

Idachaba, F. S. Agricultural Research Policy in Nigeria. Research Report, no. 17. Washington: International Food Policy Institute, 1980.

IITA. The NAFPP: A New Dimension for Nigerian Agriculture. Ibadan, Nigeria: International Institute for Tropical Agriculture, 1977.

Institut d'Economie Rurale. "Rapport de Synthèse sur les Systèmes de Culture et d'Elevage dans le Context du Mali." Bamako, Mali: Institut d'Economie Rurale, 1977. (Mimeographed)

International Food Policy Research Institute. Food Needs of Developing Countries: Projections of Production and Consumption to 1990. Research Report, no. 3. Washington: IFPRI, 1977.

IRRI. Constraints to High Yields on Asian Rice Farms: An Interim Report. Los Banos, Philippines: International Rice Research Institute, 1977.

ISRA. Recherche et Développement Agricole: Les Unités Expérimentales du Sénégal. Bambey, Sénégal: Institut Sénégalais de Recherches Agricoles, 1977.

Johnson, A. W. "Individuality and Experimentation in Traditional Agriculture." Human Ecology 1 (1972):149-159.

Johnson, G. L. "Small Farms in a Changing World." Paper presented at the Farming Systems Research Symposium on Small Farms in a Changing World: Prospects for the Eighties. Manhattan, Kansas State University, 11-13 November, 1981.

Jones, M. J., and Wild, A. Soils of the West African Savanna. Technical Communication, no. 55. Harpenden, England: Commonwealth Agricultural Bureau, 1975.

Jones, W. I. "The Food Economy of Ba Dugu Djoliba, Mali." In African Food Production Systems: Cases and Theory, pp. 265-306. Edited by P. F. McLoughlin. Baltimore and London: Johns Hopkins University Press, 1970.

Jones, W. I. Planning and Economic Policy: Socialist Mali and Her Neighbors. Washington: Three Continents Press, 1976.

Jones, W. O. "The Structure of Staple Food Marketing in Nigeria as Revealed by Price Analysis." Food Research Institute Studies 8 (1968):95-123.

Junaidu, Alhaji, Wazirin Sokoto. Muhummancen Jam'a Ga Jama'armu. ABU Publication, no. 1. Zaria, Nigeria: Ahmadu Bello University, 1972.

Kafando, T. W. Les Perspectives du Développement Rural de l'Est Volta. Paris, France: IEDES, 1972.

Kassam, A. H. "In Search of Higher Yields with Mixed Cropping in Northern Nigeria: A Report on Agronomic Work." Zaria, Nigeria: Institute for Agricultural Research, Ahmadu Bello University, 1973. (Mimeographed).

Kassam, A. H.; Dagg, M.; Kowal, J. M.; and Khadr, F. "Improving Food Crop Production in the Sudan Savanna Zone of Northern Nigeria." Outlook on Agriculture 8 (1976):341-347.

Kassam, A. H.; Kowal, J. M.; Dagg, M.; and Harrison, M. N. "Maize in West Africa: And Its Potential in the Savanna Areas." World Crops 17 (1975):75-78.

Kearl, B., ed. Field Data Collection in the Social Sciences: Experiences in Africa and the Middle East. New York: Agricultural Development Council.

Khan, A. H. Ten Decades of Rural Development: Lessons from India. Rural Development Paper, no. 1. East Lansing: Department of Agricultural Economics, Michigan State University, 1978.

King, R. "Experiences in the Administration of Cooperative Credit and Marketing Societies in Northern Nigeria." Agricultural Administration 2 (1975):195-208.

King, R. "Capital, Credit and Savings in Northern Nigeria Agriculture: Questioning Conventional Wisdom." Samaru Agricultural Newsletter 18 (1976a):17-21.

King, R. Farmers' Cooperatives in Northern Nigeria. Zaria, Nigeria, and Reading, England: Institute for Agricultural Research, Ahmadu Bello University, and Department of Agricultural Economics, University of Reading, 1976b.

King, R. "Village Variety and State University." Reading, England: University of Reading, 1978. (Mimeographed)

King, S. B. "Sorghum Diseases and Their Control." In Sorghum in the Seventies, pp. 411-434. Edited by N. G. P. Rao, and L. R. House. New Delhi, India: Oxford University Press, 1972.

Kleene, P. "Notion d'Exploitation Agricole et Modernisation en Milieu Wolof Saloum." Agronomie Tropicale 31 (1976):63-82.

Kohler, J. M. Activités Agricoles et Transformations Socio-Economiques dans une Région de L'Ouest du Mossi. Ouagadougou, Upper Volta: ORSTOM, 1968.

Kohler, J. M. Activités Agricoles et Changements Sociaux dans l'Ouest Mossi. Mémoires de L'ORSTOM, no. 46. Paris, France: ORSTOM, 1971.

Kohler, J. M. Les Migrations des Mossi de l'Ouest. Travaux et Documents de l'ORSTOM, no. 18. Paris, France: ORSTOM, 1972.

Kolawole, M. I. "Economic Aspects of Private Tractor Operations in the Savanna Zone of Western Nigeria." Savanna 3 (1974):175-184.

Kowal, J. M. "The Hydrology of a Small Catchment Basin at Samaru, Nigeria. Part III." Nigerian Agricultural Journal 7 (1970a):120-133.

Kowal, J. M. "The Hydrology of a Small Catchment Basin at Samaru, Nigeria. Part IV." Nigerian Agricultural Journal 7 (1970b):134-147.

Kowal, J. M., and Kassam, A. W. Agricultural Ecology of Savanna: A Study of West Africa. Oxford, England: Clarendon Press, 1978.

Kowal, J. M., and Knabe, D. An Agro-Climatological Atlas of the Northern States of Nigeria. Zaria, Nigeria: Ahmadu Bello University, 1972.

Kowal, J. M., and Stockinger, K. R. "The Usefulness of Ridge Cultivation in Nigerian Agriculture." Journal of Soil and Water Conservation 28 (1973):136-137.

Kriesel, H. C. "Some Economic Performance Problems in the Marketing Component of Statutory Marketing Systems." In The Marketing Board System, pp. 149-165. Edited by H. M. A. Onitiri, and D. Olatunbosun. Ibadan, Nigeria: Nigerian Institute of Social and Economic Research, 1974.

Lagemann, J. Traditional African Farming Systems in Eastern

258

Nigeria. Munich, West Germany: Weltforum Verlag, 1977.
Lahuec, J. P. "Une Communauté Evolutive Mossi: Zaongho." Etudes Rurales 37 (1970):150-172.
Laurent, C. K. Problems Facing the Fertilizer Distribution Program in the Six Northern States. CSNRD/NISER, no. 27. East Lansing: Michigan State University, 1969.
Lele, U. The Design of Rural Development: Lessons from Africa. Baltimore: Johns Hopkins University Press, 1975.
Lewis, J. V. D. "Descendants and Crops: Two Poles of Production in a Malian Peasant Village." Ph.D. thesis, Yale University, 1978.
Lewis, O. "Medicine and Politics in a Mexican Village." In Health, Culture and Community: Case Studies of Public Reactions to Health Programs, pp.? Edited by B. D. Paul. New York: Russell Sage Foundation, 1955.
Lipton, M. A., and Moore, M. The Methodology of Village Studies in Less Developed Countries. IDS Discussion Paper, no. 10. Brighton, England: Institute of Development Studies, University of Sussex, 1972.
Longhurst, R. "Work, Nutrition and Child Malnutrition in Northern Nigeria." Ph.D. dissertation, University of Sussex, England, 1980.
Longhurst, R. W.; Palmer-Jones, R. W.; and Norman, D. W. "The Role of the Social Scientist in Determining Research Priorities for Agricultural Development in the Developing World." Zaria, Nigeria: Institute for Agricultural Research, Ahmadu Bello University, 1976. (Mimeographed)
Luning, H. A. An Agro-Economic Survey in Katsina Province. Kaduna, Nigeria: Government Printer, 1963.
Marchal, J. Y. "The Evolution of Agrarian Systems in Yatenga." African Environment 2 (1977) and 3 (1977):73-85.
Matlon, P. J. "The Size, Distribution, Structure and Determinants of Personal Income among Farmers in the North of Nigeria." Ph.D. thesis, Cornell University, 1977.
Matlon, P. J. Income Distribution among Farmers in Northern Nigeria: Empirical Results and Policy Implications. African Rural Economy Paper, no. 18. East Lansing: Department of Agricultural Economics, Michigan State University, 1979.
Matlon, P. J. "The Structure of Production and Rural Incomes in Northern Nigeria: Results of Three Village Case Studies." In The Political Economy of Income Distribution in Nigeria, pp. 323-372. Edited by H. Bienen, and V. P. Diejomaoh. New York: Holmes and Meier, 1981.
Matlon, P. J., and Newman, M. "Production Efficiency and Income Distribution Among Farmers in the North of Nigeria." East Lansing: Michigan State University, Department of Agricultural Economics, 1978. (Mimeographed.)
Mauss, M. The Gift. Glencoe: The Free Press, 1954.
Maymard, J. "Structures Africaines de Production et Concept d'Exploitation Agricole. Première Partie. Un Example de Terroir Africain: Les Confins Diolamanding aux Bords du Sonngrongron." Cahiers ORSTOM (Série Biologie) 24 (1974):27-64.
McDonald, D. Asperigillus Flavus on Groundnut and Its Control in

Nigeria. Samaru Research Bulletin, no. 115. Zaria, Nigeria: Institute for Agricultural Research, Ahmadu Bello University, 1967.

McDonald, D. Trials with Copper-Sulphur Dust for Control of Cercospora Leaf-Spots of Groundnuts in Nigeria. Samaru Miscellaneous Paper, no. 45. Zaria, Nigeria: Institute for Agricultural Research, Ahmadu Bello University, 1973.

Meillassoux, C. "Development or Exploitation: Is the Sahel Famine Good Business?" Review of African Political Economy 1 (1974):27-33.

Mellor, J. W. The New Economics of Growth. Ithaca: Cornell University Press, 1976.

Mijindadi, N. B. "Staff Organization for Agricultural Planning: The Case of Nigeria." Agricultural Administration 3 (1976):239-247.

Mijindadi, N. B. "Production Efficiency on Farms in Northern Nigeria." Ph.D. dissertation, Cornell University, 1980.

Milleville, P. "Enquête sur les Facteurs de la Production Arachidienne dans Trois Terroirs de Moyenne Casamance." Cahiers ORSTROM (Série Biologie) 24 (1974):65-99.

Milleville, P. ORSTOM, Ouagadougou, Upper Volta. Interview, September 1978.

Monnier, J.; Diagne, A.; Sow, D.; and Sow, Y. Le Travail dans l'Exploitation Agricole Sénégalaise. Bambey, Sénégal: CNRA, 1974.

Morel, R., and Quantin, P. "Observations Sur l'Evolution à Long Terme de la Fertilité des Sols Cultivés à Gimari (République Centrafricaine)". Agronomie Tropicale 27 (1972):667-739.

Moreno, R. L., and Saunders, J. J. A Farming System Research Approach for Small Farms of Central America. Turrialba, Costa Rica: CATIE, 1978.

Mortimore. M. J., and Wilson, J. Land and People in Kano Close-Settled Zone. Occasional Paper, no. 1. Zaria, Nigeria: Department of Geography, Ahmadu Bello University, 1965.

Murphy, J., and Sprey, L. H. The Volta Valley Authority: Socio-Economic Evaluation of a Resettlement Project in Upper Volta. West Lafayette: Department of Agricultural Economics, Purdue University, 1980.

Netting, R. "Household Organization and Intensive Agriculture: The Kofyar Case." Africa 35 (1965):422-429.

Niang, M. Experimental Units, Kaolack, Senegal. Interview, September 1978.

Nicolas, G. "Un Village Haoussa de la République du Niger: Tussao Haoussa." Cahiers d'Outre-Mer 13 (1960):421-450.

Niger, Ministère de l'Economie Rurale. Enquête Agricole par Sondage, 1972-73. Niamey, Niger: Section Statistiques Agricoles, Direction de l'Agriculture, 1973.

Norman, D. W. An Economic Study of Three Villages in Zaria Province: I. Land and Labour Relationships. Samaru Miscellaneous Paper, no. 19. Zaria, Nigeria. Institute for Agricultural Research, Ahmadu Bello University, 1967.

Norman, D. W. "Initiating Change in Traditional Agriculture." Proceedings of the Agricultural Society of Nigeria 7 (1970):6-14.

Norman, D. W. An Economic Study of Three Villages in Zaria
    Province: 2. Input-Output Study, 2 parts. Samaru
    Miscellaneous Paper, nos. 37 and 38. Zaria, Nigeria:
    Institute for Agricultural Research, Ahmadu Bello University,
    1972.
Norman, D. W. Interdisciplinary Research on Rural Development. OLC
    Paper, no. 4. Washington: Overseas Liaison Committee,
    American Council on Education, 1973a.
Norman, D. W. Methodology and Problems of Farm Management
    Investigations: the Experience of RERU in Northern Nigeria.
    African Rural Employment Paper, no. 8. Zaria, Nigeria:
    Institute for Agricultural Research, Ahmadu Bello University,
    1973b.
Norman, D. W. "Rationalising Mixed Cropping under Indigenous
    Conditions: The Example of Northern Nigeria." Journal of
    Development Studies 11 (1974):3-21.
Norman, D. W. "The Social Scientist in Farming Systems Research."
    Paper presented at Workshop on Farming Systems Research in
    Mali, Institut d'Economie, Bamako, Mali, 15-20 November, 1976.
Norman, D. W. "Problems Associated with the Gathering of Technical,
    Economic and Social Data at the Farm Level in West Africa."
    Paper presented at the International Conference on the Economic
    Development of Sahelian Countries, Centre of Research on
    Economic Development, University of Montreal, Canada, 13-14
    October, 1977.
Norman, D. W. "Agriculture: Progress or Catastrophe in Africa?"
    Africa Report 26 (1981):4-8.
Norman, D. W., and Palmer-Jones, R. W. "Economic Methodology for
    Assessing Cropping Systems." In Cropping Systems Research and
    Development for the Asian Rice Farmer, pp. 241-260. Edited by
    the International Rice Research Institute. Los Banos,
    Philippines: IRRI, 1977.
Norman, D. W., and Simmons, E. B. "Determination of Relevant
    Research Priorities for Farm Development in West Africa." In
    Factors of Agricultural Growth in West Africa, pp. 42-48.
    Edited by I. M. Ofori. Legon, Ghana: Institute of Social
    Science and Economic Research, University of Ghana, 1973.
Norman, D. W.; Hayward, J. A.; and Hallam, H. R. "An Assessment of
    Cotton Growing Recommendations as Applied by Nigerian Farmers."
    Cotton Growing Review 51 (1974):266-280.
Norman, D. W.; Hayward, J. A.; and Hallam, H. R. "Factors Affecting
    Cotton Yields Obtained by Nigerian Farmers." Cotton Growing
    Review 52 (1975):30-37.
Norman, D. W.; Newman, D. W.; and Ouedraogo, I. Farm and Village
    Production Systems in the Semi-Arid Tropics of West Africa.
    Research Bulletin, no. 4, vol. 1. Hyderabad, India: ICRISAT,
    1981.
Norman, D. W.; Pryor, D. H.; and Gibbs, C. J. N. Technical Change
    and the Small Farmer in Hausaland, Northern Nigeria. African
    Rural Economy Paper, no. 21. East Lansing: Department of
    Agricultural Economics, Michigan State University, 1979.
Norman, D. W.; Beeden, P.; Kroeker, W. J.; Pryor, D. H.; Hays, H.
    M.; and Huizinga, B. The Production Feasibility of Improved
    Sole Crop Maize Production Technology for the Small-Scale

261

Farmer in the Northern Guinea Savanna Zone of Nigeria. Samaru
Miscellaneous Paper, no. 59. Zaria, Nigeria: Institute for
Agricultural Research, Ahmadu Bello University, 1976a.
Norman, D. W.; Beeden, P.; Kroeker, W. J.; Pryor, D. H.; Huizinga,
B.; and Hays, H. M. The Feasibility of Improved Sole Crop
Sorghum Production Technology for the Small-Scale Farmer in the
Northern Guinea Savanna Zone of Nigeria. Samaru Miscellaneous
Paper, no. 60. Zaria, Nigeria: Institute for Agricultural
Research, Ahmadu Bello University, 1976b.
Nyerere, J. K. Ujamaa: Essays on Socialism. Dar Es Salaam:
Oxford University Press, 1971.
Ogborn, J. The Control of Striga Hermontheca in Peasant Farming.
Samaru Research Bulletin, no. 207. Zaria, Nigeria: Institute
for Agricultural Research, Ahmadu Bello University, 1974.
Ogunfowora, O., and Norman, D. W. "An Optimization Model for
Evaluating the Stability of Sole Cropping and Intercropping
Systems under Changing Resource and Technological Levels."
Bulletin of Rural Economics and Sociology 8 (1973):77-96.
Olayide, S. O.; Olatunbosun, D.; Idusogie, E. O.; and Abiagom, J. D.
A Quantitative Analysis of Food Requirements, Supplies and
Demands in Nigeria, 1968-1985. Lagos, Nigeria: Federal
Department of Agriculture, 1972.
Olayide, S. O.; Ogunfowora, O.; and Essang, S. M. "Effects of
Marketing Board Prices on the Nigerian Economy: A Systems
Simulation Experiment." Journal of Agricultural Economics 25
(1974):289-309.
Olukosi, J. O. Kwara State Farm Institute Programme. Samaru
Miscellaneous Paper, no. 57. Zaria, Nigeria: Institute for
Agricultural Research, Ahmadu Bello University, 1976.
Oluwusanmi, H. A. Agriculture and Nigerian Economic Development.
Ibadan, Nigeria: Oxford University Press, 1966.
Onitiri, H. M. A., and Olatunbosun, D. eds. The Marketing Board
System. Ibadan, Nigeria: Nigerian Institute for Social and
Economic Research, 1974.
Orewa, S. I. "The Economics of Dry Season Tomato Production under
the Cadbury Tomato Processing Scheme in the Zaria Area." M.Sc.
thesis, Ahmadu Bello University, Nigeria, 1978.
ORSTOM. Enquête sur les Mouvements de Population à Partir du Pays
Mossi. 3 vols. Paris, France: ORSTOM, 1975.
Palmer-Jones, R. W. "Peasant Differentiation in Rural Hausaland."
Zaria, Nigeria: Institute for Agricultural Research, Ahmadu
Bello University, 1978. (Mimeographed.)
Peacock, J. M. The Report of the Gambia Ox Ploughing Survey. Wye,
England: Wye College, 1967.
Pedler, F. J. Economic Geography of West Africa. London, England:
Longmans, 1955.
Peil, M. "Unemployment in Banjul: The Farming/Tourist Tradeoff."
Manpower and Unemployment Research 10 (1977):25-29.
Pelissier, P. Les Paysans de Sénégal. Les Civilisations Agraires
du Cayor à la Casamance. St. Yrieix, France: Imprimerie
Fabreque, 1966.
Pendleton, J. B. "Allocative and Technical Efficiency of
Traditional Agriculture in Northern Nigeria." M.S. thesis,
Kansas State University, 1980.

262

Perrin, R. K.; Winkelmann, D. L.; Moscardi, E. R.; and Anderson, J. R. From Agronomic Data to Farmer Recommendations: An Economics Training Manual. Information Training Bulletin, no. 27. Londres, Mexico: Centro Internacional de Mejoramiento de Maiz y Trigo, 1980.

Phillips, J. Agriculture and Ecology in Africa. London, England: Faber and Faber, 1959.

Piault, C. Contribution à l'Etude de la Vie Quotidienne de la Femme Mawri. Etudes Nigériennes, no. 10. Niamey, Niger: Institut de Recherche en Sciences Humaines, 1965.

Platt, B. S. "Food and its Production." In The Development of Tropical and Subtropical Countries, pp. 96-115. Edited by A. L. Banks. London, England: Edward Arnold, 1954.

Polanyi, K. The Great Transformation. Boston: Beacon Hill, 1964.

Poleman, T. T., and Freebairn, D. K., eds. Food, Population, and Employment. New York: Praeger, 1973.

Poulain, J. F., and Tourte, R. "Effects of Deep Preparation of Dry Soil on Yields from Millet and Sorghum to Which Nitrogen Fertilizers Have Been Added (Sandy Soil from a Dry Tropical Area)." African Soils 15 (1970):553-586.

Purvis, M. J. A Study of the Economics of Tractor Use in Oyo Division of Western State. CSNRD/NISER, no. 17. East Lansing, Michigan State University, 1968.

Quinn, J. G. "Environment and the Establishment of an Industrial Tomoto Crop in Northern Nigeria." Paper presented at EUCARPIA Meeting of the Tomato Working Group, International Centre for Advanced Mediterranean Agronomic Studies, Bari, Italy, 26-30 August, 1974.

Raay, H. G. T. van. Rural Planning in a Savanna Region. Rotterdam, Holland: Rotterdam University Press, 1975.

Raay, J. G. T. van, and Leeuw, P. N. de. "The Importance of Crop Residues as Fodder." Tijdschrift Voor Economische en Sociale Geografie 60 (1970):137-147.

Raheja, A. K. "A Report on the Insect Pest Complex on Grain Legumes in Northern Nigeria." In First IITA Grain Legume Improvement Workshop, 29 October - 2 November, 1973, pp. 295-303. Edited by the International Institute for Tropical Agriculture. Ibadan, Nigeria: IITA, 1974.

Raheja, A. K. "Assessment of Losses Caused by Insect Pests to Cowpeas in Northern Nigeria." PANS 22 (1976):229-233.

Ramond, D.; Fall, M.; and Diop, T. M. Programme Moyen Terme Sahel: Main-d'Oeuvre et Moyens de Production en Terre, Matériel et Cheptel de Traction des Terroirs de Got-Ndiamsil Sesséne-Labaye (Enquête 1975). Bambey, Sénégal: CNRA, 1976.

Ravault, F. "Kanel, l'Exode Rural dans un Village de la Vallée du Sénégal." Cahiers d'Outre-Mer 17 (1964):58-80.

Raynaut, C. La Circulation Marchande des Céréales et les Mécanismes d'Inegalite Economique: Le Cas d'une Communauté Villageoise Haoussa. Cahiers des Centres d'Etudes et de Recherches Ethnologiques, no. 2. Bordeaux, France: Université de Bordeaux, 1973.

Raynaut, C. "Transformation du Système de Production et Inégalité Economique: Le Cas d'un Village Haoussa." Canadian Journal of African Studies 10 (1976):279-306.

Reboul, D. Structures Agraires et Problèmes du Développement au Sénégal. Série Travaux de Recherche, no. 17. Paris, France: INRA, 1972.

Remy, G. Enquête sur les Mouvements de Population du Pays Mossi. Rapport de Synthèse, Fascicules I et II. Ouagadougou, Upper Volta: ORSTOM, 1977.

Roch, J. "Les Migrations Economiques de Saison Sèche en Bassin Arachidien Sénégalais." Cahiers ORSTOM (Série Sciences Humaines) 12 (1976):55-80.

Rocheteau, G. "Pionniers Mourides au Sénégal: Colonisation des Terres Neuves et Transformations d'une Economie Paysanne." Cahiers ORSTOM (Série Sciences Humaines) 12 (1975):19-53.

Roth, M. J. "The Significance, Variability, and Determinants of Labor in West African Small Farm Systems: A Case Study of Eight Zaria Farmers." M.S. thesis, Kansas State University, 1979.

Rowland, M. G.; Paul, A.; Prentice, A. M.; Muller, E.; Hutton, M.; Barrell, A. R.; and Whitehead, R. G. "Seasonality and the Growth of Infants in a Gambian Village." In Seasonal Dimensions to Rural Poverty, pp. 164-174. Edited by R. Chambers, R. Longhurst, and A. Pacey. London, England: Pinter, 1981.

Saint, W. S., and Coward, E. W. "Agriculture and Behavioral Science: Emerging Orientations." Science 197 (1977):733-737.

Schildkrout, E. "Women's and Children's Work in Urban Kano." In The Anthropology of Work, Association of Social Anthropologists Monograph, no. 19. Edited by S. Wallman. London, England: Academic Press. (In Press.)

Schultz, T. W. "A Framework for Land Economics - the Long View." Journal of Farm Economics 33 (1951):204-215.

Schultz, T. W. Transforming Traditional Agriculture. New Haven: Yale University Press, 1964.

Schultz, T. W. "Nobel Lecture: The Economics of Being Poor." Journal of Political Economy 88 (1980):639-651.

Shaner, W. W.; Philipp, P. F.; and Schmehl, W. R. Farming Systems Research and Development. 2 vols. Tuscon: Consortium for International Development, 1981.

Simmons, E. B. "Planning for Agricultural Development." The Nigerian Journal of Public Affairs 1 (1971):88-102.

Simmons, E. B. Calorie and Protein Intakes in Three Villages in Northern Zaria Province, May 1970 - July 1971. Samaru Miscellaneous Paper, no. 55. Zaria, Nigeria: Institute for Agricultural Research, Ahmadu Bello University, 1976a.

Simmons, E. B. Economic Research on Women in Rural Development in Northern Nigeria. OLC Paper, no. 10. Washington: Overseas Liaison Committee, American Council on Education, 1976b.

Simmons, E. B. Rural Household Expenditures in Three Villages of Northern Zaria Province. Samaru Miscellaneous Paper, No. 56. Zaria, Nigeria: Institute for Agricultural Research, Ahmadu Bello University, 1976c.

Simmons, E. B. "A Case Study in Food Production, Sale and Distribution." In Seasonal Dimensions to Rural Poverty, pp. 73-79. Edited by R. Chambers, R. Longhurst, and A. Pacey. London, England: Pinter, 1981.

Sjo, J. B. Department of Agricultural Economics, Kansas State University. Interview, September 1979.

Smith, M. G. "Exchange and Marketing among the Hausa." In Markets in Africa, pp. 299-334. Edited by P. Bohannan, and G. Dalton. Evanston: Northwestern University Press, 1962.

Songre, A. "La Emigracion en Masa de Alto Volta: Realides y Effectos." Revista Internacional de Trabajo 87 (1973):231-245.

Steedman, C.; Daves, T. E.; Johnson, M. D.; and Sutter, J. W. Mali: Agriculture Sector Assessment. Ann Arbor: Center for Research on Economic Development, University of Michigan, 1976.

Storm, R. "Government-Cooperative Groundnut Marketing in Senegal and Gambia." Journal of Rural Cooperation 5 (1977):29-42.

Stubbings, A. D. M. Provisional Data on Crop Enterprises, Yields, Total Production and Gross Output for Season 1977-78. Funtua, Nigeria: Evaluation Unit, Funtua Agricultural Development Project, IBRD, 1978.

Sutter, J. W. Interim Progress Report of the Sahel Project of the African-American Scholar's Council. Washington, D. C.: African-American Scholar's Council, 1977.

Swanson, R. A. Gourmantche Agriculture. Ouagadougou, Upper Volta: United States Agency for International Development, 1978.

Technical Advisory Committee. Farming Systems Research at the International Agricultural Research Centers. Washington: Technical Advisory Committee, Consultative Group on International Agricultural Research, 1978.

Tiffen, M. Changing Patterns of Farming in Gombe Emirate, North Eastern State, Nigeria. Samaru Miscellaneous Paper, no. 32. Zaria, Nigeria: Ahmadu Bello University, 1971.

Titiloye, M., and Ismail, A. A. "A Survey of the Trends and Problems in the Domestic Arrangements for the Marketing of Groundnuts and Cotton." In The Marketing Board System, pp. 135-148. Edited by H. M. A. Onitiri, and D. Olantunbosun. Ibadan, Nigeria: Nigerian Institute of Social and Economic Research, 1974.

Traore, N., and Toure, M. Le Machinisme Agricole au Mali. Journeés d'Etudes Technico-Économiques sur le Tracteur Swazi Tiukabi. Promotion de la Construction Grâce à la Coopération entre les Pays en Développement d'Afrique. Bamako, Mali: CEPI et DMA, 1978.

Uchendu, V. C. "Some Issues in African Land Tenure." Tropical Agriculture 44 (1967):91-101.

Unité d'Evaluation. Evaluation de l'Opération Arachide et Cultures Vivrières. Résultats d'une Enquête Descriptive de la zone d' Intervention de l'OACV in 1976. Bamako, Mali: Unité d'Evaluation, Institut d'Economie Rurale, 1976.

Unité d'Evaluation. Evaluation de l'Opération Arachide et Cultures Vivrières. Etude Africa-Economique de 32 Exploitations Agricoles en Zone OACV. (Résultats Definitifs de l'Enquête suivi d'Exploitation 1976/77). Bamako, Mali: Unité d'Evaluation, Institut d'Economie Rurale, 1978.

USAID. Small Farmer Credit: Analytical Paper. Washington, D. C.: United States Agency for International Development, 1973.

USAID. The Gambia Mixed Farming and Resource Management Project. Washington, D. C.: United States Government, 1978.

Valdes, A.; Scobie, G.; and Dillon, G., eds. Economics and the Design of Small-Farmer Technology. Ames: Iowa State University, 1979.

Venema, L. B. The Wolof of Saloum: Social Structure and Rural Development in Senegal. Wageningen: Centre for Agriculture Publishing and Documentation, 1978.

Vercambre, M. Unites Experimentales du Sine-Saloum: Revenus et Depenses dans Deux Carres Wolofs. Bambey, Senegal: Centre National de Recherches Agronomiques, 1974.

Verneuil, P. Comment Orienter l'Investigation en Milieu Rural Africain a Partir de la Relation entre Echange Inegal, Developpement Inegal et Transfert des Valeurs. AMIRA, no. 22. Paris, France: Association Francaise des Instituts de Recherche pour le Developpement, 1978.

Vigo, A. H. S. "A Survey of Agricultural Credit in the Northern Region of Nigeria." Kaduna, Nigeria: Ministry of Agriculture, 1965. (Mimeographed.)

Watts, M. "The Sociology of Political Economy of Seasonal Food Shortage: Some Thoughts on Hausaland." Paper presented at Conference on Seasonal Dimensions to Rural Poverty, Institute of Development Studies, University of Sussex, Brighton, England, 3-6 July, 1978.

Weber, G. "Cost of Tractors in the Agricultural Services Division." Jos, Nigeria: Ministry of Natural Resources, 1971. (Mimeographed.)

Weil, P. M. "The Introduction of the Ox Plow in Central Gambia." In African Food Production Systems, pp. 229-263. Edited by R. F. M. McLoughlin. Baltimore: Johns Hopkins University Press, 1970.

Weil, P. M. "Wet Rice, Women and Adaptation in the Gambia." Rural Africana 19 (1973):20-29.

Wilcock, D. C. Michigan State University, Eastern ORD, Ouagadougou, Upper Volta. Interview, September 1978.

Wilde, J. C. de. Experiences with Agricultural Development in Tropical Africa. 2 vols. Baltimore: Johns Hopkins University Press, 1967.

Winkelmann, D., and Moscardi, E. "Aiming Agricultural Research at the Needs of Farmers." Paper presented at the Seminar on Socio-Economic Aspects of Agricultural Research in Developing Countries, Santiago, Chile, 7-11 May, 1979.

World Bank. Nigerian Agricultural Sector Review, 1979. Washington: World Bank, 1979.

World Bank. Accelerated Development in Sub-Saharan Africa: An Agenda for Action. Washington: The World Bank, 1981a.

World Bank. World Development Report 1981. New York: Oxford University Press, 1981b.

Zalla, T. A Proposed Structure for the Medium-Term Credit Program in the Eastern ORD of Upper Volta. African Rural Economy Working Paper, no. 10. East Lansing: Michigan State University, 1976.

Zandstra, H. G. "A Cropping Systems Research Methodology for Agricultural Development Projects." Paper presented at World Bank Headquarter Seminar, Washington, D. C., 1 August, 1979.

Zandstra, H. G. "Cropping Systems Research for the Asian Rice

Farmer." Agricultural Systems 4 (1979b):135-153.

Zandstra, H.; Swanberg, K.; Zulberti, C.; and Nestel, B. Caqueza: Living Rural Development. Ottawa: International Development Research Centre, 1979.

Zandstra, H. G.; Price, E. C.; Litsinger, J. A.; and Morris, R. A. A Methodology for On-Farm Cropping Systems Research. Los Banos, Philippines: International Rice Research Institute, 1981.

Author Index

Subject Index

social differentiation, 72-74
Soil
  erosion, 41, 53
  fertility 20, 45, 50, 55-56,
    186-187. See also
    Fertilizer
  types, 43-45
Sole cropping, 54, 177, 197,
  205, 213
  See also Mixed cropping
Sorghum. See Crops
Sprayers, 212
Status, 73, 105, 150 (table),
  151, 180, 192 (table)
  See also Income
Storage, 48, 81, 83, 85, 90, 91,
  112, 141 (table), 143-144,
  193-194
Striga (S. hermontheca and
  S. senegalensis), 47
Studies
  basic, 9-10, 93-99
  change, 10, 204-223
Sudan ecological zone, 31, 43
  166
  See also Zones, agroecological
Sugarcane. See Crops
Sulfur. See Fertilizer,
  inorganic
Support systems, 75, 177, 205,
  209, 213-214, 217, 220,
  227-229
  See also Credit; Markets,
    input
Surveys. See Consumption
  survey; Farm-management
  survey; Marketing/storage
  survey; Women's occupation
  survey
  See also Farming Systems
    Research; Sample selection

Technology analysis. See
  Farming Systems Research,
  testing stage
Technology, relevant improved.
  See Appropriate technology
Temperature, 43

Testing, on-farm. See Farming
  Systems Research, testing
  stage
Tractor Hire Unit (THU), 79,
  189
Tractors, 189
Transhumance, 48, 64, 168
Transport, 82
  costs, 85, 89, 107
Trypanosomiasis, 48, 168
  See also Diseases, animal

Upland. See Land, gona
Urbanization, 64

Variables versus parameters,
  23, 31, 239, 242
Village head, 66
Village studies. See Farm
  management survey

Wage rates, 106
  See also Labor, hired
Water management, 40, 50, 56
Weaver birds (Quelea quelea),
  47
Weeding, 119, 182-183, 196
  See also Labor constraint
Weeds. See Striga
West African savanna, 37, 40
Women, 65, 68, 69, 70, 105,
  107, 128-129, 151. See
  also Female gender roles;
  Female headed households;
  Labor, female; Purdah
Women's occupation survey, 97
World Bank, 75, 182, 243

Yield gap, 9, 57-58
Yields
  mixed cropping, 53, 54
  sole cropping, 53, 54, 57
Youth, 69, 73, 111-112, 127

Zones, agroecological, 31,
  37, 38 (map), 39 (table),
  42 (table), 43, 57, 99,
  166